SUCCESS

NEVER SLEEPS

SONNY BECK

SUCCESS

NEVER SLEEPS

LESSONS ON LIFE AND LEADERSHIP

Written by Dan T. Miller and Maggie Ricks

My sincere thanks to the farmers,
the hardworking people who are the backbone of our communities
and our world. Your dedication is an inspiration.

And to the agricultural enthusiasts and all others reading this book,
I appreciate you joining me on this journey.
It's been a true joy to share my life's story with you.

Sonny Beck

DEDICATION

First and foremost, I give thanks to God. Without His guidance and grace, none of this would have been possible. Every open door, every hard lesson, every opportunity to grow has His fingerprints on it.

To my parents, Francis and Pauline, I owe so much. The lessons you learned shaped the lessons you passed on to me: live within your means, reinvest what you earn, and never forget how quickly things can change. The foundation of this business, and much of what I know about life and leadership, started with you.

To my wife, Glendia, my partner in every season of life, thank you. You have stood beside me through the long days and the moments when the only way forward was faith and hard work. Whatever success we've experienced, we've experienced together.

To our children, Scott (Shantel), Kim (Todd), and Tony, you are the greatest return on investment I could ever ask for. I'm proud not just of what you've accomplished, but of who you are. You've each carried forward the values that built this family and this company: integrity, faith, service, and a belief that tomorrow can always be better than today.

To our grandchildren: Corey (Erika), Chris (Kaylee), Jason (Grace), Jordyn (Woody), Jadyn, Emily (Tyler), Ryan, Holly (Dylan), Kiley, Hallie (Ethan), Hannah (Eli) , Faith, Laura, and Elsie, and our seven great-grandchildren: you are the future of this story. I look at you and see energy, ideas, and potential that go far beyond anything I could have imagined when this journey began. My hope is not that you simply preserve what's been built, but that you challenge it, strengthen it, and take it places we haven't yet dreamed of.

This book would not be complete without recognizing the incredible individuals on our Leadership Team who continue to help shape Beck's: Scott, Tom Hooper, Bethany Gremel, and Jim Zimmer. You help forge new paths, ask hard questions, and push us toward excellence. Your leadership ensures that we never settle for "that's good enough."

To the former employees and dealers who laid the foundation, thank you. You believed in something that was far from guaranteed. You worked the fields, knocked on doors, and built relationships that turned a small operation into a trusted name.

To our current employees and dealers, you are writing the next chapter every day. Your dedication to growth, innovation, and service keeps this company strong and relevant in an ever-changing industry.

And to the future employees and dealers we haven't yet met, this company will be yours to help shape. I'm excited about the ideas you'll bring, the challenges you'll tackle, and the new heights you'll help us reach.

Success never sleeps. It's built on faith, family, hard work, and people who care deeply about doing things the right way. As I reflect on where we've been, I'm grateful. As I look at where we are, I'm proud. And as I consider where we're headed, I've never been more excited!

The best chapters are still to come.

CONTENTS

INTRODUCTION

It was nearly midnight, and the parking lot outside Beck's headquarters sat empty in the darkness. The hum of semi traffic had faded, and the last workers had gone home hours ago. Like most nights at that time, every office light in the building was off except for one. Mine.

If you happened to drive by, you'd see a faint glow in the corner office. Inside, I'd probably be leaning back in my chair, feet stretched out on the desk, staring into the dark. So many people have asked me, "Sonny, when do you sleep?" They shake their heads, wondering if I ever go home.

The truth is, I do sleep, I just don't mind trading a little of it for the quiet I get at this hour. Most nights, I would go home for dinner with my family and then slip back to the office for a few hours. That time, when the world is still and my mind is free from interruptions, is when I do my best thinking. That's what I call my "feet-on-the-desk" time.

Don't mistake it for idleness. To an outsider, it might look that way; me leaning back in a chair with nothing in my hands. But the work happening in those hours is as real as anything that fills the daylight. Feet-on-the-desk time is when I dream, when I wrestle with problems, when I imagine the future and all it can be.

It isn't easy. Silence can feel like a blank sheet of paper, daring you to fill it. Most people don't have the discipline to embrace the quiet. They would rather keep busy scrolling phones and answering emails, anything to escape the discomfort of being alone with their thoughts. I understand that, and some nights I even feel it myself. There's always work left undone, and I could easily spend those hours catching up. But I also know that if I give in, I lose the chance to think beyond today's problems. Feet-on-the-desk time isn't about what didn't get done today; it's about what might be possible tomorrow.

Focus is the key. Sometimes there's one looming question on my mind, while other nights, I juggle five ideas at once. But the point is always the same: sit in the stillness and stay with the thought long enough to make progress and get clarity. Some of my biggest ideas were born in those hours.

Of course, the past has its place too. I think about what has worked, what hasn't, and what lessons are worth carrying forward as we grow the business. But I don't dwell on what might have been. I've known plenty of people who live in the past, who still brag about a high school victory or are stuck replaying missed opportunities. But that's never been me. I want to take what I've learned, use it, and move on from it to build something even better.

I always try to choose the door marked optimism, but hard work and persistence have been the keys to my success. I have a simple mindset that I've stuck with over the years: we'll find a way. That doesn't mean ignoring reality; it means facing it with energy and determination. The future isn't just something to dream about; it's something to build. Many times, it began with my feet on the desk, late at night, when the building was empty.

Leadership celebrates action, but I believe vision is found in stillness. If all you do is act without pausing to think, you could risk everything by moving in the wrong direction. Those quiet midnight hours may not look productive, but they're where I've found clarity for the biggest decisions, where I've shaped the course of my leadership and, at times, the direction of Beck's itself. There are a lot of things beyond our control, but one thing we can control is how we prepare, adapt, and how we build new ways forward. That's what I'm thinking about when my feet are on the desk.

If you were to ask me why I do it, why I sit in that empty office at midnight staring into the dark, I'd tell you this: the future doesn't just happen. We create it. One idea, one vision, and one quiet moment at a time.

So, if you see me leaning back with my feet on the desk, just know this: outside it may be dark and still, but inside my office, the future is already taking shape.

PROLOGUE

The red Farmall tractor rumbled like a giant awakened from sleep. I gripped the steering wheel, small hands clutching the worn metal. The pedals stretched out below me, impossibly far away. I reached my short legs, straining, but my feet dangled just shy of the mark. For a moment, defeat loomed.

My dad stood nearby, silent, watching. He disappeared into the barn and returned with two blocks of wood and some twine. Without a word, he fastened the blocks to the pedals, his rough hands working with purpose. When he stepped back, I tried again. My foot reached forward, further, further, and then, contact. I felt the pedal beneath me as a grin spread wide across my face. I was just five years old, and I was driving a tractor!

The heavy machinery chugged through the open gate of our family's 80-acre farm in central Indiana, its engine vibrating beneath me as I guided it across the fields. Dad watched with steady eyes as I breathed in that intoxicating blend of freedom and responsibility.

It was 1945, and my world had opened wide, changing forever.

1

BEFORE ME

Before there was Beck's Hybrids, before there were research plots, greenhouses, or buildings, there was land. And before there was land in our family's care, there were people who understood something simple and profound: if you steward what you're given, it can outlast you.

I never met my grandfather, Lawrence Beck. He passed away two years before I was born. But I've never felt like I didn't know him. His legacy is everywhere; in the fields we farm and in the values that shaped my father and, eventually, me.

Lawrence was a farmer in the truest sense of the word. He worked the ground with his hands, trusted God with the rest, and believed deeply in taking ownership, being responsible, and working hard. His goal was to build something steady that would hold its value and serve the next generation, which became the foundation on which everything else stood.

My parents, Francis and Pauline, inherited more than land from him. They inherited a way of thinking.

Dad grew up in the shadow of the Great Depression. It shaped him in ways only those who experienced it can understand. And from that experience came a lifelong philosophy: live within your means, reinvest what you earn, and never forget how quickly things can change. Mom carried those same values, but she lived them differently. Where Dad was quiet and cautious, Mom was warm and relational. She believed just as strongly in stewardship, but she also believed in people and the importance of bringing them together. Their partnership was balanced and intentional, and it showed me early on that leadership requires different strengths working together toward the same purpose.

In 1937, that purpose began to take shape in a new way.

From the start, Beck's Hybrids wasn't just about selling seed; it was about growing something that worked. Dad believed deeply in results, not promises, and that truth matters more than comfort. That mindset shaped how the business grew.

Land played a central role in that philosophy. To our family, land wasn't just acreage; it was security. It held its value, generation after generation, and it demanded responsibility in return. You worked it, improved it, and passed it on better than you found it.

Faith also played an important role. My parents didn't talk about faith constantly, but they lived it every day. We prayed. We went to church. But more importantly, I watched how they treated people. I saw honesty and generosity without expectation of return.

As a boy growing up on the farm, I didn't think in terms of legacy. I thought in terms of chores, seasons, and getting the job done. But looking back now, I can see that everything I learned about work, trust, time, and responsibility came from being part of something that started long before me. That understanding has stayed with me my entire life.

Time, I've learned, is the great equalizer. Every person gets the same number of hours in a week; what matters is how you pour them out. My parents poured theirs into faith, family, and farming. My wife, Glendia, and I have tried to do the same with our three children, Scott, Kim, and Tony, and they have as well with their kids. We never did it perfectly, but we were always intentional. And because of that, we've built something that has endured.

This business was never about one person. It was about continuity and legacy, doing your part well, and trusting the next generation to do the same. That's the lens through which I've tried to live my life and lead this company.

Everything I've done since, every decision, every risk, every opportunity, exists because of those who came before me. And it's what has made everything that came after possible.

2

LESSONS OF A FARM BOY

Growing up on the farm meant life was a world of contrasts. My days were a mix of chores and adventure, duty and freedom. Long before many of my classmates would sit behind a wheel, I was rumbling down lanes in pickups, hauling wagons of hay or grain with tractors. There was joy in the hum of engines and the wide-open fields, but there was no mistake that these machines came with real work. Feeding livestock, cleaning out stalls, and checking fences, there was always work waiting for me from dawn to dusk.

But I loved it. I loved all the things a kid would love on a farm. The barnyard smells, the feel of dirt on my hands, the sense that each day carried both freedom and responsibility; those things rooted deep inside me.

My parents stood like pillars throughout my life. My dad was a man of few words, preferring the steady silence to the clutter of chatter. "When you're talking, you ain't learning nothin'," he liked to say. I absorbed this, not as a reprimand but as a part of farm life: the quiet of cornfields, the lowing of cattle, the language of nods, gestures, and expectations. Dad never had to spell out what was right and wrong; I just simply knew.

My mom brought warmth and conversation to balance Dad's quiet ways. Around the kitchen table, the center of our family life, she listened. She coaxed answers out of my sister, Marietta, and me when we didn't know the words to explain ourselves. Once, as I neared college graduation, she leaned across that table and asked, "What drives you?" She helped me realize that I thrived when thinking things through, turning problems over in my mind until a path appeared, and then moving forward with reckless abandon until the challenge was overcome. Where dad showed his values through silence and example, Mom led us with questions and conversation.

As farming parents, they ensured that we saw the value of effort and responsibility. And as Christian parents, they spent extensive time teaching us a set of values, morals, and a manner of living that we would carry with us far into the future.

Their partnership wasn't only felt in the home. It extended into the fields and, eventually, into a business that would shape the future. While their roles on the farm were very different, my parents modeled work and leadership in everything they did. While Dad was busy growing a crop and building a business, he never turned down the opportunity to help educate neighboring farmers on how to grow the best hybrid seed.

Mom was busy in other ways; raising two kids while taking care of the chickens, tending the garden, and driving equipment when needed. On top of that, she led the home economics clubs and played the organ at church, often finding herself in leadership roles of whatever organization she was a part of.

Dad was less sociable but equally respected, and his word meant everything. While his activities off the farm were few, he held leadership roles in the church and the Lions Club. When he committed to something, no one doubted it would be done. We lived between Arcadia and Atlanta, Indiana, just 30 miles north of Indianapolis but a world away. It was a small but close-knit community where knowing and trusting your neighbors meant everything.

As I grew older, I found joy not only at work but in play. Basketball was my passion, but I chased grades with the same competitiveness I brought to the court, pushing myself to earn A's. I was active in school and church and took on 4-H projects, tinkering with motors and electricity, eventually winning state honors that took me to Chicago and Minnesota. For a boy that was used to dirt roads and fields, those trips felt like stepping into another world.

Still, life on the farm never loosened its grip. Even as I spent time with friends and enjoyed the freedom of boyhood, I always returned each day to chores, to family, and to constant responsibility. My mom's gentle prodding and the way my dad led through quiet examples stayed with me throughout my life.

I learned it first in those days as a boy growing up on a small, 80-acre farm: work mattered, faith mattered, and the time you gave to others mattered most of all. I didn't grow up thinking about leadership. I grew up thinking about doing the job right. Long before I ever led a company, the farm taught me how to lead myself, and those lessons would carry me and our business far beyond the fields of Atlanta, Indiana.

3

MORE THAN A GAME

I grew up knowing I was part of something that started long before me. The land, the work, the expectations, they were already there when I arrived. I didn't have words for it back then, but I understood responsibility. I understood that effort mattered and that how you carried yourself counted. For me, one of the places those lessons showed up early was on the basketball court.

I've loved basketball for as long as I can remember. Growing up, it was more than just a game for me; it was a chance to be a part of something that matched the heartbeat of Indiana itself. I played through high school, then in a community league for several years after graduation. Even now, I still enjoy walking into Mackey Arena at Purdue, the electric energy of the crowd, the squeak of sneakers, the thud of the ball. Other sports caught my eye, but basketball? It was king.

On the farm, though, sports always came second to work. I always wanted to play football; the idea of it was so exciting; cool evenings spent under those Friday night lights. But when I brought it up, Dad shut it down quickly. Football overlapped with harvest, the busiest time of year. His answer was swift and firm: no. I could've sulked or argued, but I didn't. Deep down, I knew he was right. As much as I loved football, I loved the farm more. Family and farming came first.

Basketball was different. The season stretched from November to March, when the fields were quiet and work on the farm slowed. I already had a hoop set up, as most farm kids in Hamilton County did in the '50s. Long evenings were often spent shooting hoops against the weathered backboard, the sound of the ball against the barn echoing through the silence of the countryside. The thought of wearing a Jackson Central Eagles jersey was exciting to me. I worked up the courage to ask Dad if I could try out, and, unlike football, this time, he said yes. The only condition was that there would be times I would need to find my own way to practice or to games. That was fine by me. Mom agreed, and with that, I was in.

The first real lesson I took from those early years wasn't even about basketball. It was about not burning bridges. When Dad said no to football, I respected his decision. Because of that, when the opportunity to play basketball came around, he listened. That taught me something about patience, respect, and keeping relationships strong, a lesson I've carried with me all my life.

When I joined the Eagles, Coach Hugh Leeman became a central figure in my life. He wasn't just our coach; he was also my biology teacher. A Navy veteran and Butler University graduate who had played under the legendary Tony Hinkle, Coach Leeman carried himself with quiet strength. He cared about more than just the scoreboard. He watched our grades, our behavior, and our choices off the court. If something seemed off, he'd talk to our parents. Thankfully, he never had to call mine.

Coach Leeman's generosity was legendary. He'd pile half the team into his Oldsmobile, seven of us squeezed shoulder-to-shoulder, for a ride home after games. He looked after us like we were his own, and cared about us first as young men, and second as basketball players. He was always watching out for us and our futures.

On the court, his leadership stood out. He believed in finding each player's strengths and putting them in the right spot. This was a principle I'd later recognize in business as "getting the right people on the bus, and then putting them in the right seats," a concept written by author Jim Collins. It was a concept we still implement today at

Beck's when we look to find the right employees to fill roles. But back then, it was instinct. Coach kept his cool during games. If a player messed up, he didn't scream or throw clipboards. He'd talk quietly, and if the issue persisted, he benched you. Simple as that.

I'll never forget the sting of being sent to the bench. I don't recall the mistake I made, but I'll never forget the humbling walk to the sidelines. The game continued without me, and that was the lesson: no one is indispensable. Life goes on, whether you're ready or not. When I returned to the floor, I played his way, without complaint.

My role on the team (or my seat on the bus, so to speak) fit my strengths. I wasn't the tallest, but I was quick and had a good sight of the entire floor. Against the press, I'd bring the ball up the court, weaving through defenders. Other times, Coach stationed me at the free-throw line, back to the basket, like a makeshift center. I'd pass to whoever was hot or slip past a slower defender for a quick score. My size didn't matter nearly as much as my speed, awareness, and decisiveness. Basketball showed me that talent comes in many forms.

There were highlights. I was named to the all-county team as a senior, and captained the team my last year. But that final season also taught me about setbacks. Sitting out with an injury for four games was excruciating. My teammates and coach told me they missed me, which meant more than any award, but it was hard to watch the action from the bench.

Dad never let me get too proud. After one game against Western High School, where I scored 22 of our 43 points, I rode home expecting praise. He kept his eyes on the road, snowy fields passing us as we drove along. I asked him what he thought of the game and in a low voice he said, "Your team lost." That was just his way: always reminding me it was about the team, not the individual. It stung, but it stuck.

My last game is etched into memory. It was 1958, and we faced Noblesville in an afternoon tournament. The Millers had two outstanding guards, and I was still nursing my injury. Tension was thick in the gym as the lead shifted with every basket. With less than a minute left, I got the ball at the free-throw line, cut off a screen, and shot a 10 foot jumpshot. The ball fell, and the crowd erupted. For a moment, it felt like victory. But seconds later, a Noblesville player launched a desperate shot from deep. It swished as the buzzer sounded. Just like that, my career was over. Dad didn't say a word afterward. He didn't have to.

Looking back, I see what those years gave me: lessons in humility, leadership, and decision-making under pressure. My years on the court taught me to respect others, to put the team above myself, and to never assume the game (or life) revolves around one person. Even today, my wife, Glendia, teases me about being hyper-competitive. She's right, but not in the way most think. My competition is with myself, always striving to be better than yesterday.

Long before I ever led a company, I learned on that court how to be part of something bigger than myself. And while I may not step onto the court anymore, the lessons still echo in my life and give me the drive to push forward, no matter what the scoreboard says.

4

PURDUE AND THE POWER OF POSSIBILITY

I can still feel the heavy, humid summer air as I walked out of Jackson Central High School for the last time, diploma in hand. There was no big party, no balloons or celebration – but there was also no question about my future, no big decisions weighing on my shoulders. Where many of my friends and classmates debated factory jobs or sticking close to the farm, my path forward was paved long before I graduated. College wasn't optional; it was expected. And I was headed to Purdue University.

Education for farming had changed drastically since Dad's days, and while he never earned a formal degree, he used to talk about Purdue's Winter Short Course like it was Wonka's Golden Ticket. It was the early days of the Great Depression, and those December and January winters added an extra layer of misery for farm families. The eight-week Winter Short Courses or "Farmers Institute" were the best form of education for aspiring farmers. They also served as a lifeline to civilization for their families during the brutal Indiana winters.

Farmers across the state would drive through snow and ice, piling into chilly rooms packed with worn coats and hopeful hearts. For dad, it wasn't just about learning the latest in cattle and hog-raising or soil conservation; it was about hope. A break from isolation. A rare moment of connection and clarity during a time when most people didn't know if they'd be able to feed their families the following week or keep their land another year. The courses gave him a reason to believe that better days were still ahead.

Dad always said those courses gave him a leg up on other farmers, and it was clear how much they impacted him and the new practices we implemented. It gave him a place to hear new ways of doing things and a place to get the clear-cut answers he needed for farming. And in a time when nothing was guaranteed, finding any advantage to improve his farming practices was a blessing.

It wasn't just that he trusted the University and its professors; for Dad, Purdue felt like part of the community. Situated a little less than 60 miles northwest of Atlanta, Indiana, it pretty much was. A few years later, he and Grandpa jumped at the opportunity when Purdue's botany department offered samples of a new innovation called hybrid seed corn to Indiana farmers. With three acres each, they planted their first hybrid seed corn crop in 1937, and the rest, as they say, is history.

By the time I arrived at Purdue in the fall of 1958, it was more than just an academic choice. It felt like a rite of passage. The campus felt enormous to a small-town farm kid like me. Red brick buildings stretched taller than any barn I'd ever climbed, and the sidewalks hummed with the movement of thousands of students. A new sense of freedom was blossoming inside me as I stepped into a picturesque college scene that was worlds away from the dusty barns and quiet country roads of my childhood.

I wasn't the smartest person in any room, but I knew one thing for sure: no one would outwork me. I leaned on what I had: grit, incredible recall, and a stubborn belief that effort could carry you farther than talent alone. I never saw myself in competition with others. I was competing against the guy I was yesterday. That mindset had worked for me on the basketball court in high school, and it worked just as well in the classroom at Purdue.

I majored in agronomy and immersed myself in everything the College of Agriculture had to offer. But I didn't just bury myself in books. I played trombone in the marching band, joined FarmHouse Fraternity, stayed active in 4-H, served as the vice president of the Indiana FFA chapter and the vice president of the Inter Fraternity Council, and even served as an officer in several other church and student organizations. I saw those positions not as accolades but as opportunities to serve and learn, and that was important to me.

I was always willing to try new things, to push myself in ways I hadn't even dreamed of. And that's how I, Sonny Beck, a jeans and boots-wearing small-town boy from Atlanta, Indiana, landed my first role as a male model. Yes, you read that right.

At that time, LS Ayres, a fancy department store in Indianapolis, was looking for "authentic young people," and I guess a corn-fed farm boy from central Indiana fit the bill. I did interviews with a few of their executives, and before I knew it, I was walking runways in suits and sport coats, switching outfits backstage like I knew what I was doing. I still laugh thinking about it.

There was something grounding about that time in my life. Between classes, extracurriculars, and working various jobs in the summers and helping back at the home farm, I kept busy. Purdue opened my world, but it never changed who I was at the core.

One of the most meaningful experiences I remember was during my master's program. A professor asked me what I planned to do after graduation. When I told him I was intrigued by the seed business and going back home to help grow it, he looked at me and said, "Consolidation is inevitable; the big boys will swallow you up."

And while his prediction was correct about the seed industry in general, it was wrong for what I thought we could do. His comment stuck with me and helped me see the larger forces shaping agriculture, forces that I'd eventually face head-on.

PURDUE AND THE POWER OF POSSIBILITY

I graduated from Purdue in 1962, receiving the GA Ross Award for Top Male Graduating Senior. I remember a local reporter interviewing me shortly after graduating, and he asked, "What's next for you?" Without pause, I said, "I'm going back to my family farm to raise hybrid seed corn."

He nodded politely, maybe expecting something grander. But for me, there wasn't a bigger dream. I couldn't wait to get back home, where faith, family, and hard work were all that mattered.

Purdue University opened my eyes to what was possible, exposing me to new ideas and expectations that gave me the confidence to think differently, even when the path forward wasn't clear. I didn't come home with all the answers, but I came home believing that improvement was always possible.

That belief shaped how I approached every decision that followed and became the foundation for a lifetime of innovation.

The Restored Red Farmall Tractor Sonny Beck First Drove in 1945 Now Sits in Beck's History Museum at Their Headquarters in Atlanta, IN.

The Original Beck Family Homestead Purchased By Lawrence and Anna Beck (1901)

Wedding Photo of Sonny's Grandparents, Lawrence and Anna Beck (1901)

Family Portrait - Lawrence, Francis, Anna, and Dorthia Beck (1914)

Wedding Photo of Sonny's Parents, Pauline and Francis Beck (1933)

Francis Beck With Daughter, Marietta Beck, Holding Sonny Beck (1940)

Francis, Sonny, and Marietta Before Church (1942)

Family Portrait – Marietta, Pauline, Sonny, and Francis Beck (1943)

Pauline Beck in Her Garden (She Loved Gardening and Flowers)

Pauline, Sonny, Marietta, and Francis Beck in Front of Beck's First Delivery Truck

Family Portrait – Sonny and Glendia Beck With Their Three Children, Scott, Kim, and Tony

Sonny Beck on His Pony, Rocket, Holding His Dog

Sonny Beck in a Wagon on the Farm

Young Sonny Beck

Marietta and Sonny Beck Sledding With Their Dog

Sonny, Pauline, and Marietta Beck

Sonny Beck and His Dog, Skippy (1945)

Sonny, Francis, and Marietta Beck Dressed for Church

1953 Farm Bureau Officers

Pictured are officers of the Hamilton County Farm Bureau. Left to right: Francis Beck, secretary-treasurer; Walter Stewart, president; Mrs. Ralph Musselman, Sr., social and educational leader; and Paul Lindley, vice president.

Francis Beck (Far Left) and the Rest of the 1953 Hamilton County Farm Bureau Officers. Francis Served as the Secretary-Treasurer.

Francis and Pauline Beck Standing in Front of Their New Home (1959)

Pauline Beck and Gertrude (Repass) Mosbaugh (Glendia Beck's Mother) Serving on the Hamilton County Council in 1940 - Pauline and Gertrude Were Both Pregnant With Sonny and Glendia in This Photo

Sonny Beck (#44) and the Jackson Central Basketball Team

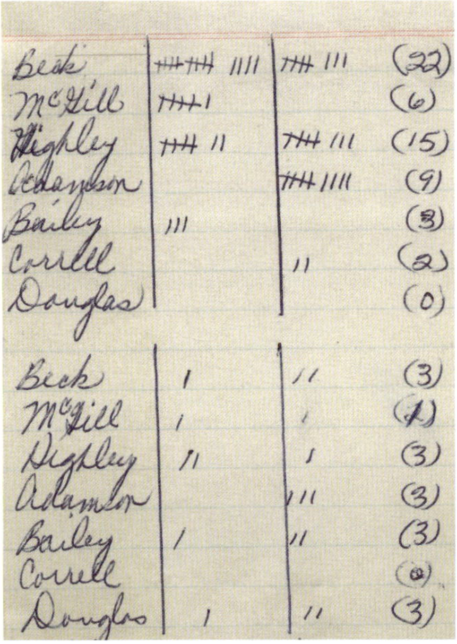

Score Sheet From The Eagle's Game Against Western High School When Sonny Scored 22 Points

Sonny Beck in a Game Against Carmel High School

"Red-Hot" Jackson Central Eagles Hope To Pull A Few Tourney Upsets

Sonny Beck (#44) and the Jackson Central Basketball Team

Sonny Beck During Warmups Before a Basketball Game

Francis Beck (Third Row, Third From Left) At Purdue's Winter Short Course Meeting That Inspired New Innovation and Practices for Beck's Hybrids

Francis Beck Receiving Purdue's Winter Short Course Outstanding Alumnus Award (1979)

Sonny Beck, Graduating Valedictorian of Jackson Central High School (1958)

Purdue University Agricultural Hall

Sonny Beck and Members of FarmHouse Fraternity

Sonny in His Purdue Marching Band Uniform – He Played Trombone

Farming, Fashion Modeling Easy For Sonny Beck

Farming and fashion modeling may seem a unique combination to many, but to Lawrence (Sonny) Beck, Atlanta 4-Her, it is part of his daily routine for eight of the nine days of the Indiana State Fair.

Each year L. S. Ayres and Co. selects by interview from state 4-H members a boy and a girl to model in their two daily fashion shows (except Sunday). This year, Beck and Judy Stewart, of Charleston, were chosen to step into the fashion spotlight.

Sonny, a Purdue sophomore, has specialized in beef projects for the past nine years and this time his 4-H program included corn, soybeans, wheat, oats, hay, tractor maintenance, junior leader, garden, poultry and electric projects.

After the fair activities subside, Sonny will return to Purdue as a sophomore to continue his studies in agronomy (soils and crops). He is the son of Mr. and Mrs. Francis Beck.

Sonny Modeling for L.S. Ayres Department Store

V. C. FREEMAN (left), associate dean of Purdue University's school of agriculture, presents the 1962 senior medal to Lawrence C. Beck, Atlanta, Hamilton county. The Purdue Ag Alumni Association awards the ag senior medal for outstanding scholarship and participation in extracurricular activities. Beck is first in scholarship in a class of 225 senior students.

Sonny Receiving the Ag Senior Medal for Outstanding Scholarship and Participation in Extracurricular Activities at Purdue University (1962)

Sonny Beck on Graduation Day From Purdue University Pictured With President Fred Hovde (Left) and G.A. Ross (Right) (1962)

Beck Selected Outstanding '62 Purdue Senior

Lawrence C. Beck, of Atlanta, Purdue agriculture senior who was graduated Sunday, was named 1962 winner of the G. A. Ross Award as the outstanding senior man in this year's 3,300-member graduating class at Purdue University.

The cash award, this year $100, has been given the last four years for academic achievement, leadership, character and overall contributions to Purdue.

It was established by George A. Ross, a 1916 Purdue graduate and the first executive secretary of the Purdue Alumni Association, four years ago with a $3,000 grant. The award consists of the annual income of the trust.

Lawrence C. Beck

Beck, who graduated with a 5.74 index, out of a possible 6.0 average for his four years, was selected by a special committee composed of representatives of the Gimlet and Reamer Clubs; the dean of women; the dean of men; representatives of Associated Women Students and Gold Peppers, and the Junior Class president.

He is a member of Farm-House fraternity, and his many college activities included Inter-fraternity Council vice-president; junior officer in the Office of Student Services; director in the Office of Faculty Committee and Academic Affairs; president of his fraternity; member of the Purdue Marching Band; vice president of the Purdue Ag Council; member of Phi Eta Sigma and Sigma Delta Chi, vice president of the State Future Farmers of America Association, and a freshman leadership camp counselor.

In addition he was named the outstanding agriculture freshman, is a member of Ceres, ag honorary, and was active in the National 4-H Club Congress and Conference. He also was chroniclor for Alpha Zeta, honorary fraternity.

Beck is a graduate of Jackson Central High School, and his parents are Mr. and Mrs. Francis R. Beck, R. 2, Atlanta.

Return Clip

Noblesville Paper

5

THE WEIGHT OF A FEW WORDS

It's funny how a few words can stick with you. Like his father, Lawrence, my dad was a quiet man not prone to long speeches or talking just to hear his own voice. But maybe the best thing a parent can hope for is that their children carry their words with them in life, and I certainly have carried his on my journey.

Looking back, the words that made the biggest impact on me were always part of a story; little moments that seemed small at the time but grew to mean significantly more with each life experience.

Like the one late-fall weekend I was home from Purdue, just in time for harvest. Dad and I had spent the day out in the fields, working sunup to sundown. Dust clung to our shirts. Our boots scuffed up the gravel as we walked side by side toward the house, the last traces of sunlight stretching long behind us.

A thought had popped into my head, and I couldn't help but ask him, "Dad," I asked, glancing sideways at him, "Why do you drive a Cadillac?" He didn't flinch or get angry. Didn't ask what kind of smart-mouthed college student he'd raised. He just kept walking, hands tucked into his worn pockets.

A few steps later, he stopped and looked at me. "I work very hard at my job on the farm every day to save up every penny," he said, "so that occasionally, I can buy something I really want that makes me happy."

That was it. And it was everything. He didn't try to justify it. He didn't need to.

He and Mom had always lived simply. Because they were children of the Great Depression, they made sure we understood the value of a hard-earned dollar. Pennies mattered. But here was this quiet truth from my dad: joy had a place in life, too. If you work hard every day, it's okay to enjoy the fruits of that labor once in a while.

He wasn't a man who lived to indulge. He was a man who believed in balance, working hard, and finding small moments to enjoy what that work made possible.

Many years later, I heard a quote from Zig Ziglar that reminded me a lot of my dad's philosophy: "You can have everything in life you want if you will just help enough other people get what they want." Like my dad, Ziglar believed that true success is found not in self-centered ambition but in helping others thrive. I like to think that I, too, have carried that mentality into everything we do at Beck's. By focusing on genuine efforts to help farmers succeed and achieve their goals, we create a ripple effect of positivity that eventually comes back to us.

Another lesson came wrapped in an even shorter sentence. One that took me years to fully unpack.

He told me once, "You have to trust someone, or you can't play marbles with them." At the time, it seemed like a quaint line about a kids' game (marbles isn't exactly the cornerstone of adulthood), but it stuck with me.

He used to talk about how marbles was the game of trust. The boys would kneel in a circle, each with their favorite "shooter" marble, aiming to knock marbles out of the ring. It ran on honesty. You had to count on the other kids to call it straight and admit when their marble had been hit. If someone cheated, the whole game fell apart.

Years later, I came to realize that marbles had a lot more to teach me about life than I initially realized. At Beck's, trust is everything. Whether it's an employee, a customer, or a vendor partner, you can't run a business while checking every move. You need honesty and integrity. After all, if someone cuts a corner once, chances are they'll do it again.

Even in euchre, a game I still enjoy playing today, you've got to trust the honor of the other individuals at the table. You can't catch every wink or ear tug. You play knowing that you're counting on someone else to be straight with you.

I never knew marbles and card games could form the basis of a leadership philosophy. But, over time, I came to realize that it was one of the best ones I have ever learned. The lessons of trust, honesty, and integrity that those childhood games taught me were simply an extension of the core values instilled in me from my earliest days.

My parents believed in a simple code: make good choices. They didn't pile on rules. Outside of one, be home before midnight; they just expected us to do the right thing. But the few expectations they had, they meant something.

One summer, after another year at Purdue, I was back home helping on the farm. I can recall a night I stayed out too late. I was a grown man in college, after all, and responsible for my own time while I was on campus. It was well past midnight when I walked through the back door, and my dad heard me, but said nothing.

The next morning, I got out of bed around six, put on my boots, and headed for the tractor, ready to spend the day in the fields. The cultivator was already hooked up, waiting. Dad met me at the barn. He looked at me, not angry, just steady, and said, "Wait a minute. There isn't going to be anybody running my equipment on that little sleep. You go do something else today."

He turned and walked away, and I felt awful.

He didn't punish me. Didn't raise his voice. He just told me, with that one quiet sentence, that I wasn't fit to do the job right that day. He was protecting me and the work, and it was about safety, responsibility, and respect. He never brought it up again, but I never forgot it.

Today, I see so many parents struggling to protect their kids from a world filled with distractions. I don't blame them, but I still believe that trusting a child to make good choices and holding them accountable when they don't is one of the best lessons you can give.

Years ago, I taped a small piece of paper to my desk. On it was a verse from Psalm 133:1:

"How good and pleasant it is when God's people live together in unity!"

That verse reminded me of how lucky I was to work with people who share my values in business, in life, and in faith. It's a simple truth, but like so many things my dad said, it goes deep. You have to trust people, and you have to get along. That doesn't mean agreeing all the time, but it does mean showing up with honesty, humility, and a steady hand just like he did.

My dad may not have said much. But when he did, I listened.

When someone you love says something worth remembering, write it down. Let it settle in, because those words will stay with you. And one day, they might just guide you when you need them most.

6

WHATEVER YOU MEASURE

A quote we attribute to how Dad thought and lived is, "Whatever you measure tends to improve."

He lived by that, and I suppose I do too. Whether it was a stand of corn or the way we ran the business, Dad believed that keeping track of results wasn't about numbers; it was about truth. You can't know what's working unless you're willing to measure it. That simple idea became a guiding light for how we farmed and how we built the business.

When I came back home in 1964, fresh from Purdue and full of ideas, I didn't just want to go to work every day; I wanted to grow something bigger. I'd been thinking about how farmers could share what they were learning, how we could build on one another's knowledge and experiences instead of keeping them siloed. So that February, I organized something simple: a Ham and Bean Supper.

That first meeting wasn't anything fancy. We had 40 people show up: neighbors, friends, men who'd been farming twice as long as I'd been alive. We didn't talk about hybrids or try to push sales. It was just farmers helping farmers, swapping ideas about seed and soil and what the next season might bring.

I didn't know it then, but that supper would turn into what we now call our PFR Insight Meetings. It was the start of something that went beyond selling seed. It was all about helping farmers succeed.

That fall, we took another leap and hosted our first-ever Field Day, which is known today as our annual Becknology Days. We had about

50 visitors that first year, which pales in comparison to the over 20,000 farmers we welcome to our various facilities in recent years. Those hot August days were a chance for farmers to walk plots, ask questions, and talk about what was working… and what wasn't. But behind all that was a bigger dream I couldn't shake: what if we could measure those variables, test those products, and figure out what truly made a difference, so our customers didn't have to?

That's where the idea for Practical Farm Research, or "PFR," came from. In college, I dreamed about experimenting with different methods for growing corn. Yes, the seed mattered, but I wanted to look at all the other management factors that could affect yield; the things we could control if we only took the time to measure them.

Back then, land-grant universities did a lot of that kind of research. But over time, funding dried up and priorities shifted. I kept thinking, "If they won't do it, someone has to." Farmers needed unbiased information they could trust. They needed data that would help them make better decisions without gambling their livelihoods on trial and error. So, we started doing the testing ourselves.

By the mid-to-late 1970s, we got serious about it. We launched studies on population and planting dates. Then we added studies on nitrogen rates, fertilizer trials, and tillage practices. We tested everything we could think of that might make a difference. Most farmers wanted to try those things themselves but couldn't afford to take the risk. So, we took it for them.

Before that point, we were really a production company that happened to sell the products we produced. But PFR helped us think differently, and eventually we started shifting to a marketing company that happens to produce most of what we sell. PFR became our way of proving to farmers that our purpose wasn't just to sell product, but to help them improve their operations, no matter what name was on the bag.

We started publishing our annual PFR Book, and the program took on a life of its own. We poured time, money, and manpower behind it because we believed in it. Farmers deserved real data they could use, and if that meant sharing results that didn't make us a dime, so be it.

That commitment was the foundation, but to truly serve farmers across the region, the program needed to grow far beyond a single location.

In 1987, my oldest son Scott officially joined the business, and by the early 1990s, he was leading the PFR program. Under his direction, it grew from a single research site into a network of nine dedicated locations. Today, we conduct research on more than 1,200 acres, running over 150 studies and evaluating hundreds of products and practices every year.

We invest millions annually in PFR because we still believe what my Dad taught me: if you measure it, you can improve it. And what we measure helps farmers improve their profitability; that's what makes it worth every dollar.

One thing I'm especially proud of is that our PFR research focuses more on testing products we don't sell than on those we do. It's completely unbiased. We publish the results exactly as they are, good or bad, and share that data with over 70,000 farmers across the country. Our goal isn't to sell more seed; it's to help farmers make better decisions. If they trust us enough to buy from us because of that, then we've done something right.

But of course, sticking to your principles isn't always easy. I'll never forget the years between 2014 and 2016 when the economy took a downturn. Like many businesses, we were tightening our belts, and our Leadership Team started asking tough questions.

We found ourselves debating whether it was time we cut either PFR or our longstanding 100% Free Replant Policy, an offering that furnishes free replant seed to growers who experience an unsatisfactory stand. Both programs were expensive. Cutting one would've saved us a lot of money. I remember sitting there, looking at the numbers in front of me, thinking about what those programs really meant. PFR wasn't just research; it was trust. And our 100% Free Replant Policy wasn't just a guarantee; it was a promise.

The truth was clear to me. I looked around that boardroom table and we all agreed, if we cut either one, we stop being who we are.

In the end, we kept both.

That decision cost us in the short run, but in the long run, it defined us. I've always believed a company's values aren't tested when things are easy; they're tested when times get tough.

This reminded me of the late 1990s and early 2000s, when our testing data showed that, with Roundup herbicide, farmers could significantly reduce soybean seeding rates if they used seed treatment and a good herbicide like Roundup to control weeds.

Publishing that meant telling farmers to buy less seed. From a business standpoint, it made no sense. But when we looked at the numbers, our Leadership Team knew that making the recommendation would, in turn, help farmers succeed.

Over time, that honesty paid off. Those same farmers who planted fewer soybeans trusted us more. Many of them came back to buy seed for more acres or for different crops. They saw we weren't just selling to them; we were standing alongside them.

That's the kind of decisions that keep me grounded. Because at the end of the day, it's not just about being the biggest or the flashiest company. It's about keeping the trust of the people who depend on you.

In 2018, our management team went to Purdue University and met with Dr. Grey, an Ag Econ professor. During his presentation on market discipline, he mentioned that there are three qualities a great company can stand on, but you must pick one: you can have the lowest price, you can offer the best products, or you can create and thrive on the best customer relationships. We pushed back on Dr. Grey, trying to make him understand that, as a management team, we felt like Beck's did a good job at all three. But he challenged us, and made us think deeper. He told us that we had to pick one, because no single company has the resources to excel at all three. After deep reflection and careful thought, we came to a consensus: we would own customer relationships, building experiences, and helping farmers succeed. PFR is just one of the ways we do that.

It's a great responsibility and an even greater privilege to keep learning, innovating, and doing the hard work that helps farmers stay profitable. Dad's words still echo in my head: "Whatever you measure tends to improve."

We've been measuring for decades now; not just yields, but the trust our customers place in us and our willingness to stand behind them, even when it costs us. That is the true measure of our success, and it's improved more than any yield we've ever seen.

7

THE PRICE OF GROWTH

The kitchen clock ticked louder than usual that morning. I sat across from my mom, Pauline, on a cold afternoon in late December, the crisp corner of a legal pad poking out from under my elbow. I'd done the math three times, reviewing the financial books she kept in pristine order. It always came back to the same number: half a million dollars. That was how much we needed to pre-pay for fertilizer and chemicals for the coming season. That was how much cash flow we didn't have.

I saw an opportunity I knew was too good to pass up: a December 30 early-pay discount that meant significant savings and lower input costs the following spring. But Mom was squarely aligned with Dad's philosophy of not borrowing money. I didn't even need to ask her what she thought about us taking out a loan; in my heart, I knew what I wanted to do went against their beliefs.

This story is about more than money. It's about habits and values, and most importantly, it's about making room for change.

That moment with my mom (and the not-so-subtle reminder that the Beck family doesn't borrow money) didn't come out of nowhere. I grew up in a world shaped by tradition, resilience, and the hard-earned values of people who had survived situations worse than I could imagine. Those values taught me to stretch every penny and only to spend what I had. They were rooted in the collapse of the economy during the 1930s, and to my parents, especially my dad, they weren't just ideas. They were the keys to our survival.

Dad had both grown up and started his family business amid the Great

Depression. Friends and neighbors throughout Hamilton County were losing their farms; people he knew, families he'd grown up with. Loans they thought would keep them afloat sank them instead. With prices for crops and livestock at an all-time low and no new jobs in sight, banks swallowed up family farms all around us.

But Dad and Mom chose a different path. They lived frugally, canned vegetables, reused what they could, and saved every dime. They kept the farm through grit, prayer, and the absolute refusal to mortgage or borrow. And when the world turned around in the 1940s, they didn't loosen their grip. Prosperity was an illusion; it could disappear overnight. To him, it wasn't a matter of if, but when another recession or Depression would hit, and he knew that being cautious with his money would serve him well in both good and bad times alike. The Depression profoundly impacted my dad's views on life and business, muting grandiose dreams of wild success to focus on modest and steady growth.

His simple business plan called for reinvesting most of the profit back into the business, and it was very simple: Grow a little crop. Sell a little seed. Buy another little farm. Repeat.

Doing business of any kind, whether during the Great Depression or in a booming economy, calls for intelligent saving, careful planning, vigilant management, and, in Dad's opinion, an absolute refusal ever to borrow money.

I honored Dad's rule for years. But by 1967, I saw the writing on the wall. To grow our seed business, to really grow, we had to increase our capacity, and I was ready for the challenge. We struck a deal with Trojan Seed Company in Minnesota to supply them with seed they could sell at retail. The only problem was that, to fulfill our end, we needed to boost our production and harvest more seed, and we couldn't meet the demand with the equipment we had. We were already building a new seed dryer at the time, but we desperately needed another tractor and another planter, and we needed them fast.

I had assumed that, when it came to the survival and growth of our business, there would be needs that would trump my parents' long-standing attitude towards borrowing. I had assumed wrong.

When I brought the idea to Dad, he didn't get angry. He didn't need to. He just reminded me how borrowing had destroyed so many farms during the Depression, how he and Mom had survived without it. And how, from his perspective, nothing was more dangerous than debt.

Dad looked me square in the eye and said, "Around here, when you make money, son, you can spend money."

And that was that. This conservative attitude toward borrowing forced me to think harder about how I would provide solutions to sustain company growth. I walked away from our conversation with a weight on my shoulders and an idea; if I couldn't buy a tractor and planter, maybe I could build one.

At that time, we had a Uni-Harvester tractor with a corn picker that unhooked and converted into a snowplow, and it sparked a vision for a different kind of tractor-planter. I teamed up with my friend, Chuck Ritter, who worked for Trojan Seed, and we collaborated on a design. Between his local machine shop and our own farm shop, we brought it to life: a custom-built planter mounted on our existing Uni tractor, and it cost us next to nothing.

Later, a representative from a company called New Idea, based in Ohio, came by and liked what he saw. He asked me if he could replicate it, and I said yes, I didn't own a patent for it. Two years later, when I needed another planter, he gave me a great deal. That was one way to grow without borrowing. And for a while, it was enough.

Then came the 1970s. Inflation surged, interest rates skyrocketed, and banks handed out loans like candy. Farmers mortgaged everything and then land values crashed. Those who borrowed too much lost it all.

Dad's caution paid off again. We tightened our belts, avoided debt, and weathered the storm. It reinforced everything he believed.

But by 1990, Beck's hit a wall in terms of growth. We'd built something bigger than my wildest dreams, a family-run seed business with real momentum, but we couldn't keep going without investing in our future. And investing meant borrowing… not just a little; half a million dollars' worth.

This time, there was no innovative solution. No creative workaround. Just math. We needed the money to purchase the inputs by December 31 to secure a 10% discount. The problem was that our customers didn't pay us until January 2. It was a simple cash flow problem with financial consequences if we didn't act. I knew I wanted to take advantage of this opportunity, but it meant taking a risk and going against my parents' beliefs.

I was fifty years old on that December day in 1990 when I first asked my mom for permission to break one of Dad's most steadfast rules. I needed a three-day loan. She said, "Go ahead and do it. But we won't tell Dad, at least not right away."

So, I went to our local banker who'd known my family for years. I explained everything: our plan, our needs, our track record. He asked for a deed to one of our farms as collateral. I handed it over; he stuck it in a drawer and gave us the loan. We were off to the races.

Eventually, Mom and I told Dad. He wasn't happy. But he couldn't argue with the result; he saw what it meant for our company and our growth.

That moment changed me. I still respect the value of frugality. I still avoid unnecessary debt. But I'm no longer afraid to borrow - wisely, carefully, and when the future demands it.

What I learned is that tradition isn't static, and sometimes, honoring your roots means bending the branches. It doesn't mean everything changes or is subject to change; it simply means that adapting is sometimes necessary. The values my parents instilled in me were a foundation. And when it came time to build on it, I did.

We've been growing ever since.

8

THE WHITE PAPER

It was March 1992 when the phone rang in my office. I picked it up and immediately recognized the voice on the other end; Howard Doster, one of my professors from Purdue. He had been a constant presence during my college years. Before he could finish explaining why he was calling, I cut him off with a laugh.

"You're not about to ask me to write another paper, are you?" It had been years since I stepped foot in a classroom, and I thought I was done with red pens and late nights!

He laughed, that deep, familiar chuckle I remembered from years ago. The truth is, we'd always had that kind of relationship, half-teasing, half-serious. I'd taken several of his economics classes, and while I respected him immensely, I could still see those old papers of mine marked up with his edits in red ink.

But this call wasn't about another assignment. It was about something bigger. Earlier that year, Purdue had named me one of the first-ever recipients of its "Distinguished Agricultural Alumni" award. It was an honor I hadn't expected, and as part of the program, my professor wanted me to prepare a presentation.

"Sonny," he said, "I'd like you to prepare a paper to deliver as part of a panel at the awards program." He wanted me to answer one big question: What will happen to small and mid-sized seed companies in the future?

I hung up the phone and sat there, staring at the wall. It was a question

I hadn't realized I'd been wrestling with for years. I thought of another Purdue professor, Dr. Hiltz, who once told me I should consider staying in academics. His reasoning? "Small farms and small businesses like yours will all be swallowed up by big corporations." And I remembered Earl Butz, the dean of agriculture at Purdue, who had told our class bluntly: "Go big or get out." That didn't sit well with me. In fact, just about everything I'd done since graduating from Purdue was my attempt to prove them wrong.

So, when I started writing this paper in 1992, I didn't begin by collecting books and reading articles. I instead reflected on many conversations I had with my dad while growing up. He had drilled a story into me from the time I was a boy, one that had nothing to do with seeds or cornfields. He liked to talk about cars.

"Back in the 1920s and 1930s", he said, "there were nearly 80 companies making cars and trucks, many of them right here in Indiana. Kokomo. Indianapolis. Cities that today might not scream "automobile capital," but back then, the roar of engines and the clatter of assembly lines were practically in our backyard."

But by the 1960s and 1970s, I realized that those 80 companies had dwindled down to about six or eight. Most of them were national brands proving just how difficult it had become to stay viable in the marketplace. Then by the 1970s and 1980s, it narrowed again to just the big three: Ford, GM, and Chrysler.

I'd seen the same thing in classes at Purdue, in books I read, and in the world around me. Beer. Telecommunications. Airlines. It didn't matter the industry, the same pattern played out repeatedly: small companies faded away, swallowed by giants who, in time, often lost touch with the very customers who had built them.

I realized a pattern always formed: once capitalism narrowed the market to three companies, those brands stopped listening to what their customers wanted and instead focused solely on providing the most money to their shareholders. That opened the door for foreign carmakers to swoop in. They paid attention to what people actually wanted and, in turn, stole the market share.

This lesson wasn't just about cars. It was about survival in a capitalistic market, and it stuck with me. So, when writing my paper in 1992, I didn't have to bury myself in research or dig through old files. I already knew the answer: the seed business was heading down the same road as the car business. Consolidation was inevitable.

But that wasn't enough. I couldn't just stand in front of a room full of farmers and say, "Sorry, the little guy's doomed." That's not who I am. And it's not what Beck's stands for. I wanted to tell a different story. One that said survival wasn't enough, you needed to thrive. But thriving was only possible if we dared to be different.

Somewhere along the line, the paper took on a new name in our conversations around Beck's. It became known as "The White Paper," and it was more than just a document. It was a way of seeing ourselves clearly: where we stood, where the market was heading, and how we could chart our own course in an ever-changing industry. It reminded us that Beck's wasn't just about producing seed. It was about building relationships and becoming a resource for farmers. It was about listening when others stopped listening, and being the partner farmers could trust through good seasons and bad. From that day forward, everything we did had to align with that idea.

The White Paper wasn't just words on a page. It was a promise. A reminder that even in a world of giants, there's always room at the top for those willing to put the farmer first. That's what "Helping Farmers Succeed" means to us. It's who we are, and who we'll always be.

Aerial Photo of Beck's Plant (1950)

Francis Beck Pictured With An Early Cutter-Style Detasseler

Sonny Beck at 21 Years Old Bagging Corn (1961)

Francis Beck Received the Fisher's Trophy for Having the Highest Corn Yield
Three Times More Than Other Growers in the Area

A Ham and Bean Supper in the Mid 1960's - Now Known as PFR Insight Meetings

Mail this card by Thurs, ~~Feb. 2~~ JAN 30 or call Arcadia YU ~~4-2593~~ 45090
or YU 4-5135 by Fri. ~~Feb. 3~~ JAN 31 for reservations.

Check one or more:

___ Soak a handful for me, I'll be there.
___ I'd like to bring ~~my hired man or~~ a neighbor.
___ Sorry, I have to babysit.
___ ~~Sorry, I'm afraid there won't be enough ham in the beans.~~

___ Sorry, I need to visit a sick friend.
___ Sorry, I'm afraid there won't be any good jokes.
___ *Can't eat with you, but will come for the meeting at 7:30 p.m.*

Name ______________________

Mon., Feb. 3, 6:30 p.m. Lions Hall in Arcadia.

Invitation to a Beck's Ham and Bean Supper (1968)

Francis Beck Answers Customer Questions About Newly Developed Hybrids at a Beck's Field Day in 1965, Now Known as Becknology™ Days

A Beck Employee, Wilbur Lacy, Giving a Tour at An Early Beck's Field Day (1965)

Beck's First Single Cross Test Plot (1964)

Francis Beck Evaluating Hybrids by Measuring Ear Length

Sonny Beck Showcasing Record-Breaking Corn Ears (1972)

Sonny at a Beck's Field Day in 1980, Talking About Corn Hybrids

Scott Beck (Right) Shows a PFR Test Plot to Customers

Sonny and Scott Beck Celebrate the Original Breaking of the 300 Bushel Barrier in PFR (1994)

Customers on a Buggy Tour Ride at Beck's Field Day (1980)

Francis Beck Inspecting Seed Ears in the Drying Facility (1970s)

Sonny's Custom-Built Planter Mounted on an Existing Uni-Harvester Tractor (1967)

Sonny Beck's First Major High Rise Construction Project – a State-of-the-Art Seed Conditioning Tower (1970)

Sonny and Francis Beck (1979)

Sonny Beck Receiving the Purdue Distinguished Ag Alumni Award from Agricultural Dean Robert L. Thompson (1992)

Sonny Beck Presenting the "White Paper" at the Purdue Distinguished Ag Alumni Awards Program (1992)

All 11 Recipients of the 1992 Class of the Purdue Distinguished Agricultural Alumni Award (Sonny Pictured Far Left)

The Beck Family and Friends at the Purdue Distinguished Ag Alumni Awards Program in 1992. From Left to Right: Pauline, Francis, Kim, Scott, Glendia, and Sonny Beck. Back Row: Todd Marschand and Lee Rulon.

Distinguished Purdue Ag.
Panel Discussion
April 24, 1992
By Sonny Beck

WHAT WILL HAPPEN TO SMALL AND MID SIZED SEED COMPANIES IN THE FUTURE?

<u>3 Categories</u>

I. Some will make no changes.- Carry on for better or for worse just as they have been doing.

II. Some will merge:

A. As the offering price gets too high

B. or need for growth capital becomes to great

C. or no family members to carry on

III. Some will innovate, adapt, and grow.

These are the ones you will primarily see 5 and 10 years from now. Here are 3 characteristics you may find in these small to mid sized seed companies that choose to innovate, adapt, and grow.

1. Niche Marketing-

A. Whether that be marketing waxy corn, special soybeans for a particular market or simply as in our case a geographical area. In other words, find out what you do best and do more of it.

B. We classify our marketing area as a 120 mile radius around our plant. We will sell outside that area but not aggressively seek it.

C. We specifically test and develop hybrids for this area. Many

2

vill not

sease

hem

ome

III.

t

... corn for the south. We don't really want to expand down there, so we sell it to 3 other companies who market in various southern states.

B. Another of our private inbreds works so-so in Indiana, but appears to be great in Nebraska so we may market it to a Nebraska company on a royalty basis.

C. Another example is a new soybean release we will have for next spring. It comes from a major soybean breeding program (ie. 5 years of testing and thousands of tests.) The major company actually had 2 closely related lines. Until last fall they didn't know which one they wanted to release. They have 4000 Bu. of each variety. They have decided on the one with highest yield overall in the midwest (10

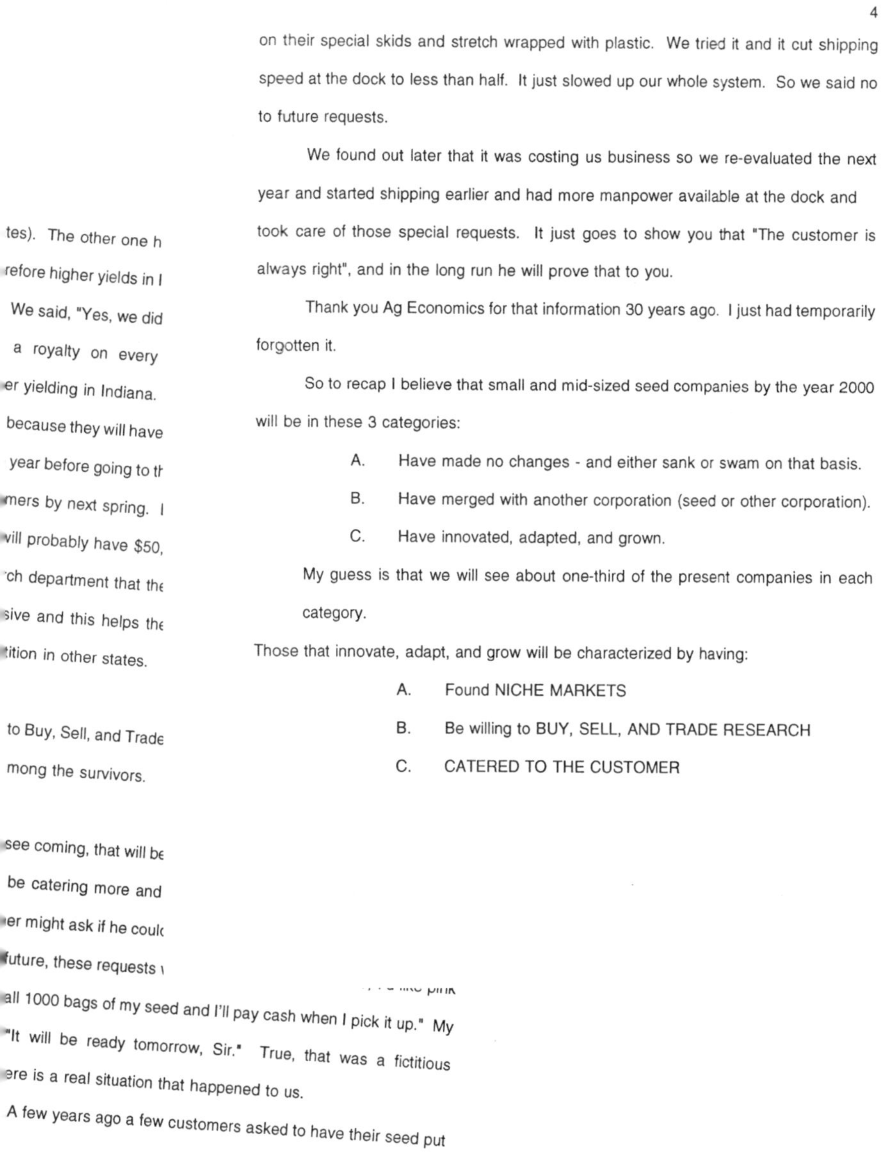
4

on their special skids and stretch wrapped with plastic. We tried it and it cut shipping speed at the dock to less than half. It just slowed up our whole system. So we said no to future requests.

We found out later that it was costing us business so we re-evaluated the next year and started shipping earlier and had more manpower available at the dock and took care of those special requests. It just goes to show you that "The customer is always right", and in the long run he will prove that to you.

Thank you Ag Economics for that information 30 years ago. I just had temporarily forgotten it.

So to recap I believe that small and mid-sized seed companies by the year 2000 will be in these 3 categories:

A. Have made no changes - and either sank or swam on that basis.

B. Have merged with another corporation (seed or other corporation).

C. Have innovated, adapted, and grown.

My guess is that we will see about one-third of the present companies in each category.

Those that innovate, adapt, and grow will be characterized by having:

A. Found NICHE MARKETS

B. Be willing to BUY, SELL, AND TRADE RESEARCH

C. CATERED TO THE CUSTOMER

Sonny Beck's "White Paper" on the Topic of "What Will Happen to Small and Mid-Size Seed Companies in the Future?" that he presented at the Purdue Agricultural Alumni Awards Program.

Sonny Beck in Front of a Record 224.6 Bu. Corn Crop Featuring BECK 90X, a Private Hybrid. Beck's Bright Gold Jacket Was Very Popular in the 70s and 80s, and is Now a Sought After Collectible (1972)

9

IF WALLS COULD TALK

The boardroom carries a certain weight for everyone who steps foot inside. Maybe it's the worn carpet and old wood table, or the hush that settles just before a meeting begins. For some, it's the intimidation factor; the windowless room where business-altering decisions are made and life-changing topics are discussed.

For a select few, it's the memories shared within those four walls. Tears have been shed. Laughter has boomed. We've celebrated, we've debated, and we've come together in the hardest of times to move this company forward.

But for me, it's what lines the walls that speak loudest. Taped with intention are pages from presentations Scott and I have delivered over the years; quotes, diagrams, and principles that have steered us through growth and change. To most, they are just thoughts memorialized in time. To me, they are visual representations of how far we've come and how close we've stayed to our roots.

And the questions that continue to echo since the '90s:

How do we grow without becoming like the big guys?
How do we expand and still stay true to who we are?

It's a weight I've carried with me ever since I pulled away from Purdue, degree in hand, and navigated the quiet country roads back home to help run the family business. I've always believed in growth. It wasn't just a dream; it was our mission. And yet, every mile we expanded, we risked losing the very soul that made us special in the first place.

Back in the 1990s, we were growing at a rate of 10-12% a year. After five decades of consistent growth, I knew we were on the edge of something great. But it was time to make a crucial decision: should we stay within a 90-mile radius of our headquarters, or expand our sales territory to 120 miles?

One night, I found myself alone with the map, staring at county lines; 90 miles felt safe, 120 felt like stepping into the unknown. "Do we need to grow to survive?" I asked myself. The answer was yes. "Do we need more seed sales to support new research and testing?" Yes, again.

That decision to expand our sales territory from 90 miles to 120 was a turning point for our business. It might not sound like much now, but at the time, it felt monumental. And in hindsight, it truly was. Adding just 30 miles to our footprint increased coverage by 78%, reshaping our entire sales trajectory. What's impressive is that, years later in 2021, we expanded into Nebraska and grew our footprint from 14 to 15 states, with almost 70 million potential corn acres. But that milestone, as significant as it was, didn't carry quite the same weight as that initial leap.

Why? Because growth isn't always about size, it's about perspective. And perspective changes everything. That first expansion wasn't just about acres or miles; it was about mindset. It forced us to get better, smarter, and more focused. It proved we could grow and stay true to ourselves.

In 1992, I wrote the "White Paper," which was introduced in the last chapter. It was titled, "What Will Happen to Small and Mid-Sized Seed Companies in the Future?" When I presented it during Purdue's Distinguished Agricultural Alumni Award celebration, I shared my thoughts on consolidation in the ag industry. At that time, I knew the seed industry had reached a point where it would likely see accelerated consolidation in the next five to ten years.

My prediction was that three types of companies would appear: those that would not change (for better or for worse), those that would merge, and those that would innovate, adapt, and grow. The third group would be the lone survivors. These insights led me to an even bigger question: "If I believe this to be true, what does it mean for the future of Beck's?"

In capitalism, consolidation always happens, and I told our team, "If we don't grow, we go." But we knew we had to grow differently. Consolidation tends to produce companies focused solely on shareholders. "Most companies forget about the customer, we can't be one of them."

And that's what we've learned: if you take care of the customer first and if you help them succeed, they will reward you. That philosophy became our anchor, and it's how we knew we'd beat consolidation. We would help farmers succeed.

That mindset led me to seek guidance from others. I've always believed in learning, knowledge, and education. I've also realized that learning doesn't stop at the classroom door. It grows in quiet moments of reading, in hard decisions, and in the shared stories of those who've been down similar roads. The quote, "You will be the same tomorrow as you are today except for the books you read and the people you meet," really hits home for me. So, I turned to books. To thinkers and leaders whose experiences stretched beyond agriculture but resonated deeply with what we were building here at Beck's.

One book helped shape my perspective on company growth: Barbarians to Bureaucrats: "The Corporate Life Cycles," by Lawrence Miller. I picked it up in the early 1990s, not long after we decided to expand.

In it, Miller parallels a company's life cycle to that of a human's: energetic and strong in youth, rigid and weak in old age. In the early years, companies thrived on innovation and passion. But as they age, they slow down and become obsessed with policies and procedures instead of responsible risk-taking and forward thinking. The very spirit that brought them success begins to fade. His words were a wake-up call.

In the late 1990s, Beck's was still lean and scrappy. We were young in a particular way: even though we had been around since 1937, we were young in staking a claim in a corporate-dominated industry that was on the verge of consolidation. Competitors were making decisions with shareholders in mind instead of farmers. It was exactly what we didn't want to become, and I saw a warning in Miller's words. A day would come when we would face the temptations of comfort and ease.

Twenty years later, his theory was the basis of my 2015 State of the Company presentation. Synergy was the "sweet spot" for Beck's, and it was crucial that for our company to grow, we needed to maintain a business model focused on long-range motives, faith, creativity, and selflessness.

Another author who shaped my outlook on business was Jim Collins, and his book, Good to Great: Why Some Companies Make the Leap and Others Don't. One quote in particular refined my philosophy on people. He said, "If a great company consistently grows revenue faster than its ability to get enough of the right people to implement that growth, it will not simply stagnate, it will FALL."

Packard's Law, he called it. And that word, "growth," was right there, staring back at me. "Packard's" referred to David Packard, founder and leader of the infamous Hewlett-Packard company. Packard attributed the success of his company to the people behind it, not the equipment or technology they sold. This theory helped me understand that growth was pointless if not backed by purpose and people. We weren't just selling seed; we were building a team, a mission, and a culture. And we could only accomplish our mission of helping farmers succeed if we, as Collins put it, had the right people on the bus and in the right seats.

That idea put language around something I'd already felt in my gut. People used to ask me, "What do you hire for?" I'd say, "Attitude and work ethic. What else do you want to know?"

And that's still true. If someone shows up with the right heart and the will to work, I'll bet I can find a job for them. And maybe it's not the job they imagined, but we'll discover something that fits. That's part of leadership, too: seeing a role where others don't. We've tried to build a company that people want to be a part of and where they're treated well. One where they're not just selling seed but living out a mission.

That's something I learned growing up under the Golden Rule. Treat others how you want to be treated. Collins helped me see that treating people well wasn't just kindness, it was strategy. It builds longevity and loyalty. And it was how we would thrive where others failed to survive.

Another voice that holds a prominent space on the boardroom wall is Simon Sinek. I've read a few of his books over the years, and he always gives me a lot to think about when reflecting on our business. In Start With Why, he writes, "People don't buy what you do, they buy why you do it."

It's something I believe wholeheartedly; no matter how great your product is, it won't matter if you don't believe in what you're doing. My passion for farming and seed started when I was a kid and hasn't let up since. That passion is my "why." And we need people at Beck's who feel the same.

Growth for us has never been about chasing numbers. It's been about chasing purpose. That's what it all comes back to. Every decision, every risk, every long day, it all comes back to one question:

Does this help farmers succeed?

If the answer's yes, we do it. If it doesn't work, we still did the right thing. That's the kind of company we want to be.

It's how we beat consolidation.
It's how we stay scrappy.
It's how we grow without losing who we are.

And if the wall in our boardroom ever starts to feel cluttered, that's fine by me. It just means we're still learning.

10

THE HEART OF CAPITALISM

I remember the sting of the crisp autumn air as it hit my face that morning. It was the fall of 1956. I was 16 years old, standing alone on the edge of 35 acres of Indiana farmland that I had just purchased with my own savings. It was a modest plot next to the 424 acres my father farmed, but it offered me a 10% share in the family business and a new perspective. As I looked out over that field, I didn't just see dirt. I saw opportunity. I saw ownership. I saw a future.

That moment was capitalism made real.

I mentioned my dad's simple business plan before. It was one rooted in reinvesting profits back into the enterprise. Grow a little crop. Sell a little seed. Buy another little farm. Repeat. It wasn't long after my first 35-acre purchase that another 160 acres went up for sale, and Dad and I jumped at the opportunity. We used our family's savings and grew what was once a small 80-acre farm to 600 acres. It was my first introduction to using one asset to build another asset, a key component of capitalism that everyone should learn.

In my experience, nothing has held its value or steadiness quite like land. It was our family's version of a bank: dependable, trustworthy, and generational. Nothing appreciates in value quite like land, and, year in and year out, it gives you something you can rely on. And for my family, with a strict no-borrowing policy, it was the only bank we trusted.

When I talk about capitalism, I'm not talking about the dog-eat-dog, negatively portrayed version you hear about in the media - the kind buried under misinformation and cynicism. I'm talking about the kind of capitalism that ties you to the land and makes your hands dirty with purpose. The kind that calls on you to take risks and gives you the freedom to dream.

At its best, capitalism gives people the opportunity to build, to fail, and to try again, and that's a beautiful, terrifying thought. For many people today, especially younger generations, the terror of failing overwhelms the promise of what could be. They're taught to view capitalism as a world where the rich get richer, and the rest just get stepped on, and I don't blame them for being wary.

But capitalism also offers something irreplaceable: the chance to live by your own hands and thrive on your own instincts. But that freedom comes with a cost and a single question: "Can you survive?" Because at its core, capitalism is about survival.

I truly believe capitalism is the best economic system in the world. It promotes freedom, creativity, responsibility, growth, and mutual relationships based on trust, dependability, and service. You can pursue your God-given talents without fear that someone or something will take them away. That's no small thing. Capitalism is part of what has made the United States a great nation. Like everything, capitalism also has negatives. But I'm with Winston Churchill, who said, "Democracy was the worst form of government except for everything else." The same thing is true for capitalism; every other system pales in comparison.

Capitalism only thrives in the presence of freedom. Without it, the whole system collapses into something unrecognizable: more government control, more regulation, and fewer choices. Before you know it, the power to shape your life and how you want to run your business is stripped away from you.

To embrace capitalism is to embrace change, which also requires the ability (and necessity) to adapt. This was the point I touched on in my White Paper in 1992. Change is a constant in capitalism, and the seed industry was a clear example of that. For decades, hundreds of small seed companies scattered the landscape. Then the winds shifted, and consolidation came. One by one, those small, mostly family-owned companies disappeared, and I refused to let that be our story.

So, we leaned in. Not by expanding recklessly, but by digging deeper into who we were. What was our purpose? Not just selling seed but helping farmers succeed. That meant offering unparalleled service that reflected our emphasis on our relationships with our customers. That clarity was empowering. It gave us a north star as the industry shifted beneath our feet. I knew we had to learn to accept change by becoming part of the change.

We had to get creative in other ways, too.

In the mid-1990s, GMOs burst onto the scene. Monsanto introduced the Roundup Ready gene, a weed control revolution, and we found ourselves in quite a predicament. We couldn't afford the labs, tools, staff, or the time to achieve that level of research.

But capitalism doesn't wait for perfect conditions. So, we pivoted. When Monsanto set out to find seed companies to license its Roundup Ready soybeans, we jumped at the opportunity. Beck's was one of 250 companies vying for the partnership that would give us access to their genetics. Monsanto's team respected the work we had done up to that point, testing their varieties, and we had lengthy discussions about our efforts to enhance seed performance.

Eventually, they chose us, along with a dozen other companies (although we were the smallest), to sell their technology, a partnership that produced BECK 302RR soybean variety. It went on to be Indiana's number one selling Roundup Ready soybean variety for three consecutive years and remained in our product lineup for over a decade. It wasn't luck. It was capitalism in action.

While corporations are a big part of our U.S. capitalist system, our mindset puts us in a unique position. Many companies focus only on their stakeholders. Incentives get tied to outcomes while leaders chase quarterly gains. They get complacent, lose sight of the people at the other end of their products, and stop innovating and changing. They get cautious and stagnant, focusing solely on protecting what has worked in the past.

At Beck's, we chose a different path.

Our capitalistic mindset prioritizes innovation and change in the name of our customers. We worry less about the bottom line and more about farmers. And because we're family-owned, we can reinvest – not to drive higher share prices, but into research. Into people. And into relationships.

It's made all the difference.

At the end of the day, capitalism isn't just an economic system. It's a stage for human stories as it profoundly influences and shapes the lives, experiences, and narratives of individuals and societies. It's the reason my family could plant roots in Indiana soil and grow something lasting. Because when capitalism works, when it's personal, purposeful, and principled, it isn't just about survival. It's about legacy.

11

THE TABLE WE SHARE

It's early Monday morning, and the sun has just started to peek over the horizon. Just like every Monday before, I gather up my coffee cup and white legal pad and walk across the hall to the boardroom. Inside, it's warm. Familiar. And around the conference table sit four people who know me better than just about anyone else.

Scott Beck. Tom Hooper. Bethany Gremel. Jim Zimmer.

We're not just coworkers. We are something deeper: a Leadership Team. But I'd argue that phrase doesn't do it justice. What we have is more like a family. Not just because we share a mission, but because we know how to debate, disagree, and stand firm for what we believe in and still walk out of the room in alignment, respecting each other. That kind of team doesn't come from job titles and tenure. It comes from years of hard-earned trust.

It comes from something I first learned not in an office, but on the basketball court. Those seasons playing basketball in my high school days taught me more about leadership than any textbook ever could. We may not have always been the team to beat, but we had heart. And when we played to our strengths and trusted in each other's talents, we won.

And before Beck's ever became the company it is today, I experienced my first Master Class in teamwork right at home from my parents. Mom, ever the social butterfly, could turn a stranger into a friend with one conversation. Dad, thoughtful and quiet, chose his words purposefully. They were opposites. But together, they were unshakeable.

I didn't know it then, but I was watching the early version of a Leadership Team built on balance, and it's a lesson that stuck with me.

And then there was Lee Rulon.

I grew up with Lee, five years my junior, as our families had been friends for as long as I can remember. Officially joining the business in 1972, Lee became the retail sales manager at Beck's Hybrids. But to me, he was more than a colleague; he was a friend, a confidant, and an incredible teammate. Smart, friendly, and a quick study, I leaned heavily on Lee's talents and charisma, especially when it came to pitching new business ideas to my dad in the 70s and 80s.

Pitching a change to Dad was always nerve-wracking. He'd listen, sure, but he always had a hundred questions. And each one was more specific than the last. Lee and I developed a routine. We'd hole up together with a pot of coffee and play out every possible reaction Dad might have. Lee focused on the sales angle. I brought the research and agronomy side. We'd rehearse. Debate. Argue. We knew that if we stumbled or weren't prepared, that would be it. Dad's answer would be "no."

And eventually, we got good. Good enough that Dad started saying yes to our ideas. But more than that, Lee and I built something irreplaceable: a team. Two guys who trusted each other with the future of the company. We didn't always agree, but we questioned each other from a position of recognizing each other's worth and reliability.

People often assume that team members can't challenge each other. I learned from Lee that this assumption isn't true. Or, to put it another way, it can't be true. Teammates with a solid foundation and relationship can and should challenge each other. Lee taught me the power of genuine trust and respect within a team, and that was a true blessing.

Years later, it was that experience that helped me shape our leadership into the team it is today. I wanted what Lee and I had: mutual respect, complementary skills, and above all, integrity. The kind you can feel in your gut.

That's what we have now.

Scott, Tom, Bethany, and Jim each came to Beck's at different times, with different stories. But they all brought something more valuable than experience. They brought character of the highest caliber. Not the kind you put on a resume. The kind you reveal when no one's watching.

It takes a special kind of person to work on a Leadership Team within a family-based company. Tom, Jim, and Bethany are genuinely some of the best people I've had the privilege of knowing, each having achieved success before coming here. Tom arrived first, a fierce and steady force. He has the sharpest memory of anyone I know, and a great mind for financial analysis and budgeting. Today, he serves as our director of business. Bethany, director of culture and brand experience, brought an innate understanding of people and purpose. She is incredibly smart and has a true gift for bringing people together. And Jim, director of supplier relations, in turn, contributed a deep understanding of the industry and a thoughtful calm that never wavered. He has a unique, down-to-earth way of applying competitor insights into company strategy.

And my son Scott, well, Scott's always been more than a team member. Scott is one of the most special people in my life, and I couldn't be prouder of him. Everyone at Beck's sees his relentless dedication, how he pours himself into everything he does for the company and for other people. He's caring and selfless; a true servant leader.

While that phrase gets tossed around too easily these days, with Scott, it's profoundly accurate. As president of Beck's, he embodies servanthood, demonstrating a unique commitment to the well-being of our Leadership Team, our employees, dealers, and our customers. He makes a special effort to break down every barrier and shine a light into every corner. It's not easy to be so selfless, but he makes it look effortless.

What makes these individuals and our Leadership Team different isn't that we always agree, because we absolutely do not. In fact, our best meetings are the ones where we debate and discuss for hours as we fill whiteboards and legal pads, meticulously weighing every major decision. But underneath all that lies an unshakeable foundation: integrity. We believe in each other's intentions, knowing that every person at that table is committed to doing what's best for Beck's.

I realize now that there is one word that encapsulates the remarkable individuals on this Leadership Team. Humble. I didn't truly grasp the meaning of "humble" for a long time. I understood what it took to be humble, but not what it meant. There's a difference between knowing a concept and truly understanding it. These four individuals embody what it means to be humble. They're quick to praise others and guide them in decision-making without claiming recognition.

Leaders often possess strong characteristics and a reliance on their own abilities. This can make it challenging for them to display genuine humility. Not a single person sitting around our leadership table operates this way. Not one. We listen more than we talk, we give credit before taking it, and we will admit when we're wrong. We lean in. Challenge each other. Laugh. Sometimes we pray. There's a current in the room: trust, respect, belief in what we're doing.

And on Monday mornings, when I take that same walk from my office to the boardroom with my coffee and my legal pad, I'm reminded of just how rare that is and how grateful I am to share this table with them.

It's why I cherish this team like family. Because in the end, we aren't just building a company. We're building a future. One where integrity leads, egos are left at the door, and trust is the foundation of everything we do. That's what a real Leadership Team looks like.

12

DECISIONS THAT DEFINE US

If you've ever stepped into my office or the boardroom at Beck's, you've probably seen a sheet of paper or a whiteboard hanging on the wall with a bold "T" drawn right down the middle. I've leaned on that simple symbol for as long as I can remember.

It looks basic, maybe even childish. A single line with a crossbar. But over the years, that T-bar has helped guide me through tough choices, opened honest debates, and brought clarity to moments of uncertainty. More than just a decision-making tool, it's become a reflection of how I try to lead: with fairness, openness, and respect for every voice in the room.

It was a strategy I learned: to evaluate how the potential positives and negatives of a decision balanced out. Over the years, there have been many decisions that members of our family and the Leadership Team have collectively weighed in on. And while I've already touched on it, when it comes to discussing the trajectory of our business, there is one major decision we had to make in the 1990s that really stands out in my mind:

"Should we sell beyond a 90-mile radius of Beck's plant?"

After five decades of careful, purposeful growth, we were at a crossroads: do we grow or do we hold steady? We all knew the stakes, but nobody wanted to say the wrong thing.

I grabbed a piece of paper and drew the big "T" on the top . On the left side, I wrote: Option 1: Stay Within 90 Miles.

On the right: Option 2: Expand to 120 Miles.

"Alright," I said, "let's see where this takes us."

If you know anything about our business, you probably know where we landed. We chose to grow and survive.

We have continued to use the T-Bar over the years to help weigh the pros and cons of monumental decisions. These discussions generally follow the same structure: first, the responses from the room are hesitant. A few people offer reasons for "yes." Then someone braves a "no." As each thought is spoken and written down, the energy in the room shifts as ideas keep flowing.

Moments like that showed me the deeper value of the T-bar. It's not just about making one choice; it's about how we make choices, together. Over the years, I've come to see why this simple tool carries such weight.

In the case of expanding to 120 miles, using a T-Bar helped us determine that the disadvantages of incurring higher costs in production and delivery per unit were more than offset by the extra income obtained through additional sales.

By starting with the "yes" side of the T-Bar, we can build momentum in a positive direction without forcing anyone into agreement. It's not about manipulation; it's about creating an atmosphere where people feel uplifted. Too much of life pulls us down, but the T-bar begins by lifting us up.

Second, every point (whether yes or no) takes a complicated issue and breaks it into something digestible. Seeing the written list grow can also spark new ideas and force us to confront assumptions that may come off as facts. After all, when you see your own words staring back at you, it's hard to ignore your blind spots. The T-Bar makes fairness unavoidable.

Lastly, the T-Bar strategy thrives in its simplicity, reinforces informed decision-making, and sparks creativity.

The T-bar may sound like a tool for groups or teams, but it matters just as much for individuals. I've used it alone, jotting down points on a legal pad, and still felt the clarity that comes from putting thoughts into two columns. But where it shines brightest is in community.

Bethany Gremel once told me that she always felt comfortable debating around me because I argued both sides of an issue. "You never shut me down," she said. "Too many leaders decide early and won't let anyone disagree. But with you, I could take a stance either way, and I knew I'd be heard."

That meant a lot to me, because at its core, the T-bar isn't really about columns of pros and cons; it's about character. It's about building a culture where every person feels respected and where objectivity is actually put into practice.

Other members of our Leadership Team have found true benefit in using the T-Bar to make important decisions.

Jim Zimmer said, "Sonny, you even convinced me to argue against myself in a T-bar discussion." That wasn't my goal, but it shows how unavoidable fairness becomes when both sides are laid out. Tom Hooper also loves to debate in this way, and my son, Scott, always works both sides of the discussion before making a decision. I believe that the younger generations of our family who work in the company have also bought into the effectiveness of this concept.

We don't dictate decisions at Beck's; we guide them. The T-bar is just one way of keeping conversations honest, fair, and open. It taps into something timeless: the human ability to listen, reflect, and weigh both sides with integrity.

It's a reminder that leadership isn't about having all the answers. It's about creating the conditions for the best answers to emerge and respecting all voices.

For me, that's what the T-bar represents. It's not just lines on a page. It's a way of living out the values that matter most.

Beck's Field of 302RR Soybeans Planted in 7.5 in. Rows (1996)

Beck's Team at the Farm Progress Show In Front of a
"Big Enough to Get the Job Done Yet Small Enough to Do it Right" Poster Highlighting
Beck's Ability to Adapt While Staying True to Our Purpose.

CORPORATE

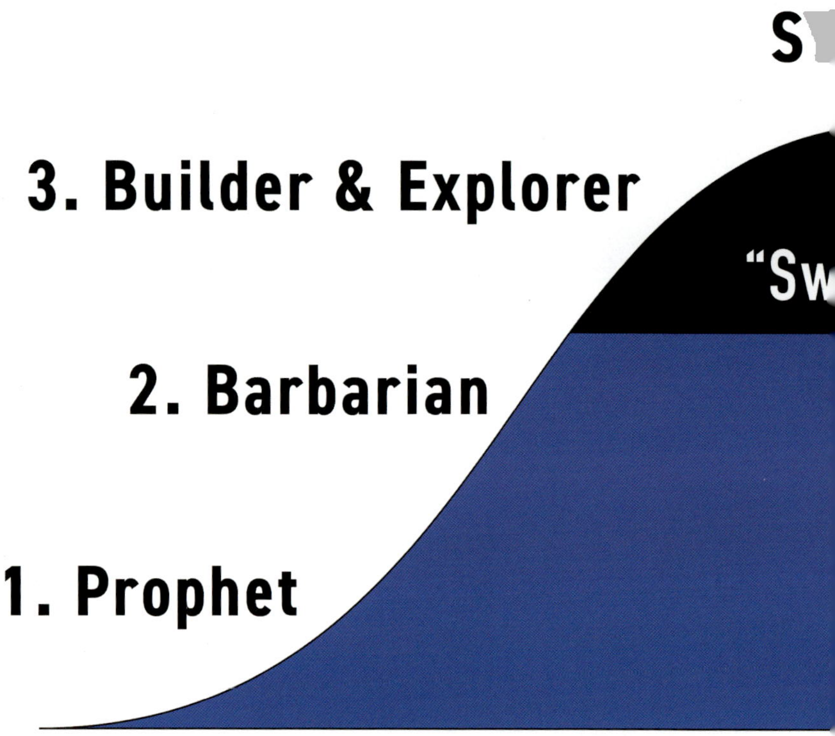

Culture When Growing (Stages 1-4)

Long Range Motives, Faith In Others, Simple Structure, Creative Leaders, Creative Problem Solving, Unselfish People, Fun

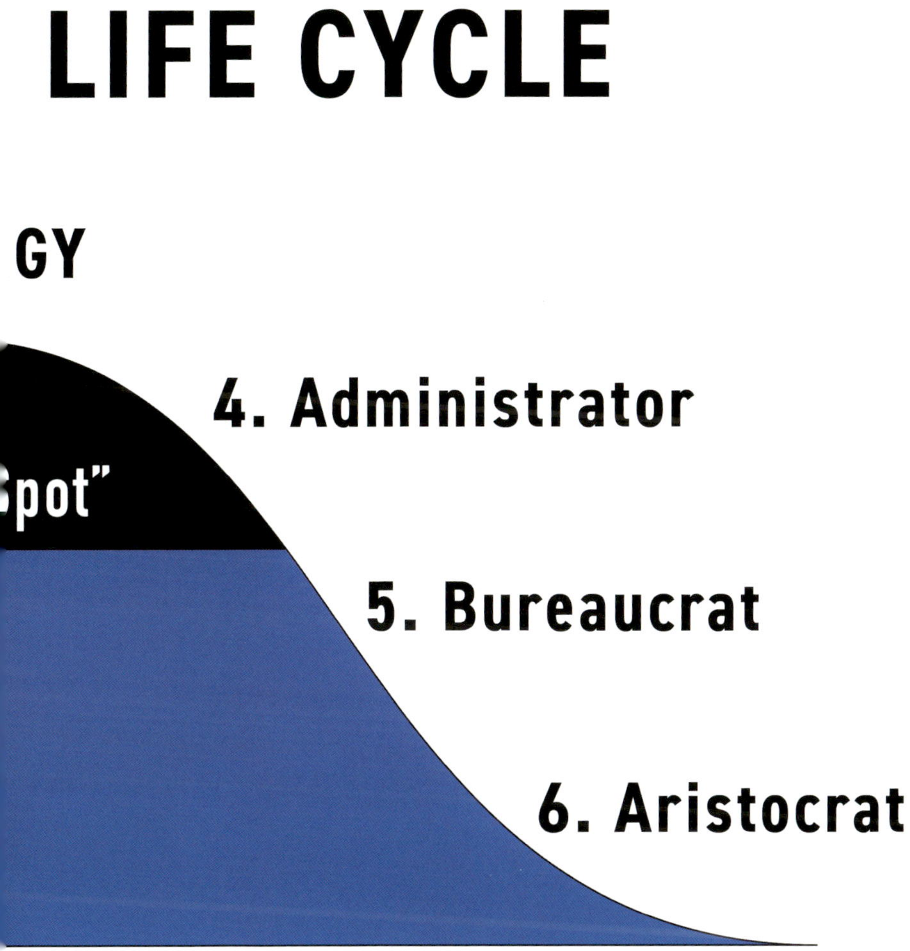

Culture When Dying (Stages 5-6)

Short Range Motives, Little Faith In Company, Complex Structure, Leaders Managing Risk, Problem Solving By The Book, Selfish People, Just A Job

Beck's "Version" of Lawrence Miller's The Corporate Life Cycle Highlighting Beck's Goal of Staying in the "Sweet Spot," and Building From the Prophet to Administrator Stage Without Ever Entering the Bureaucratic Stage.

Beck's Historical Marketing Map Highlighting The Year of Expansion Into Each State.

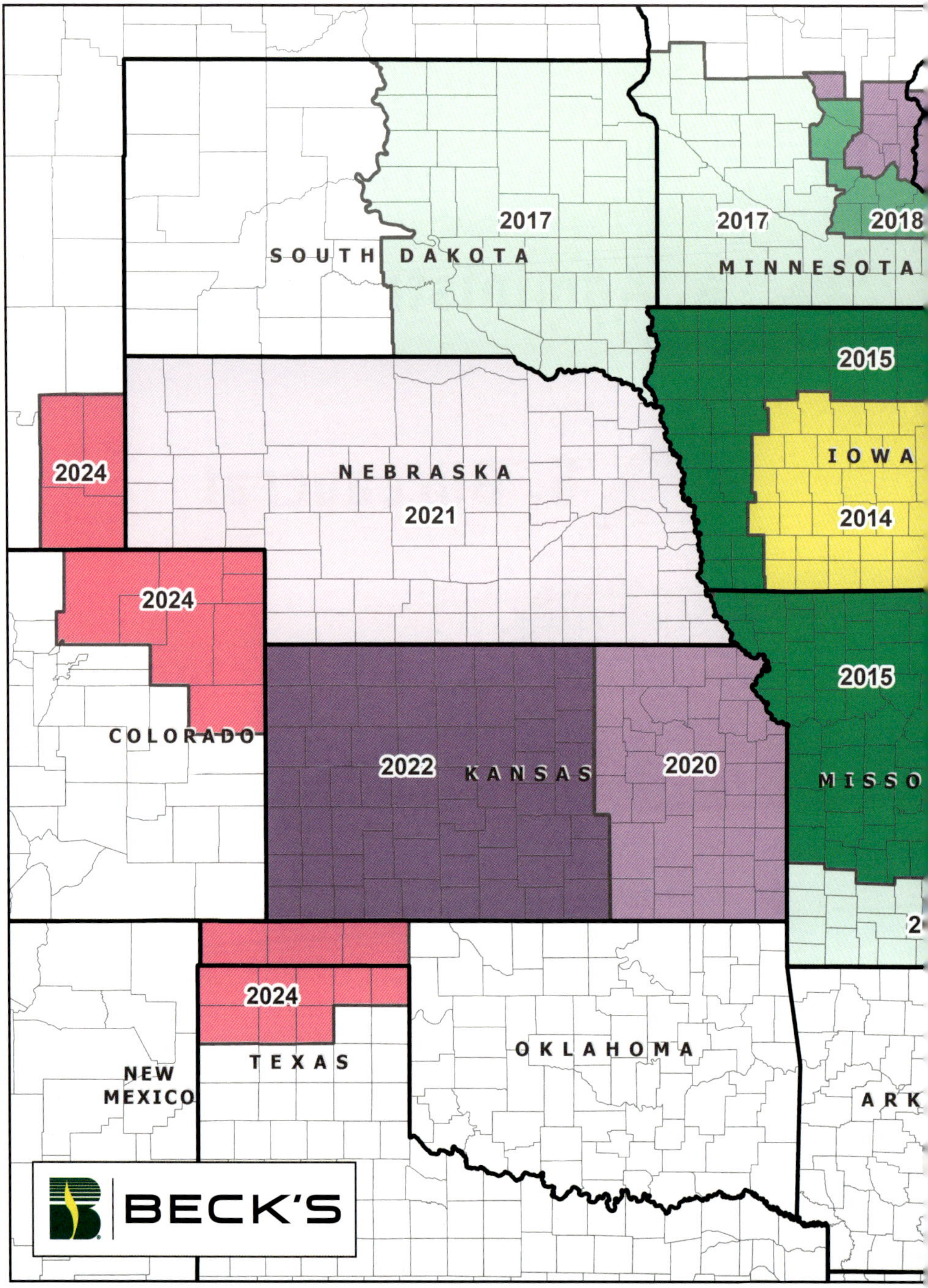

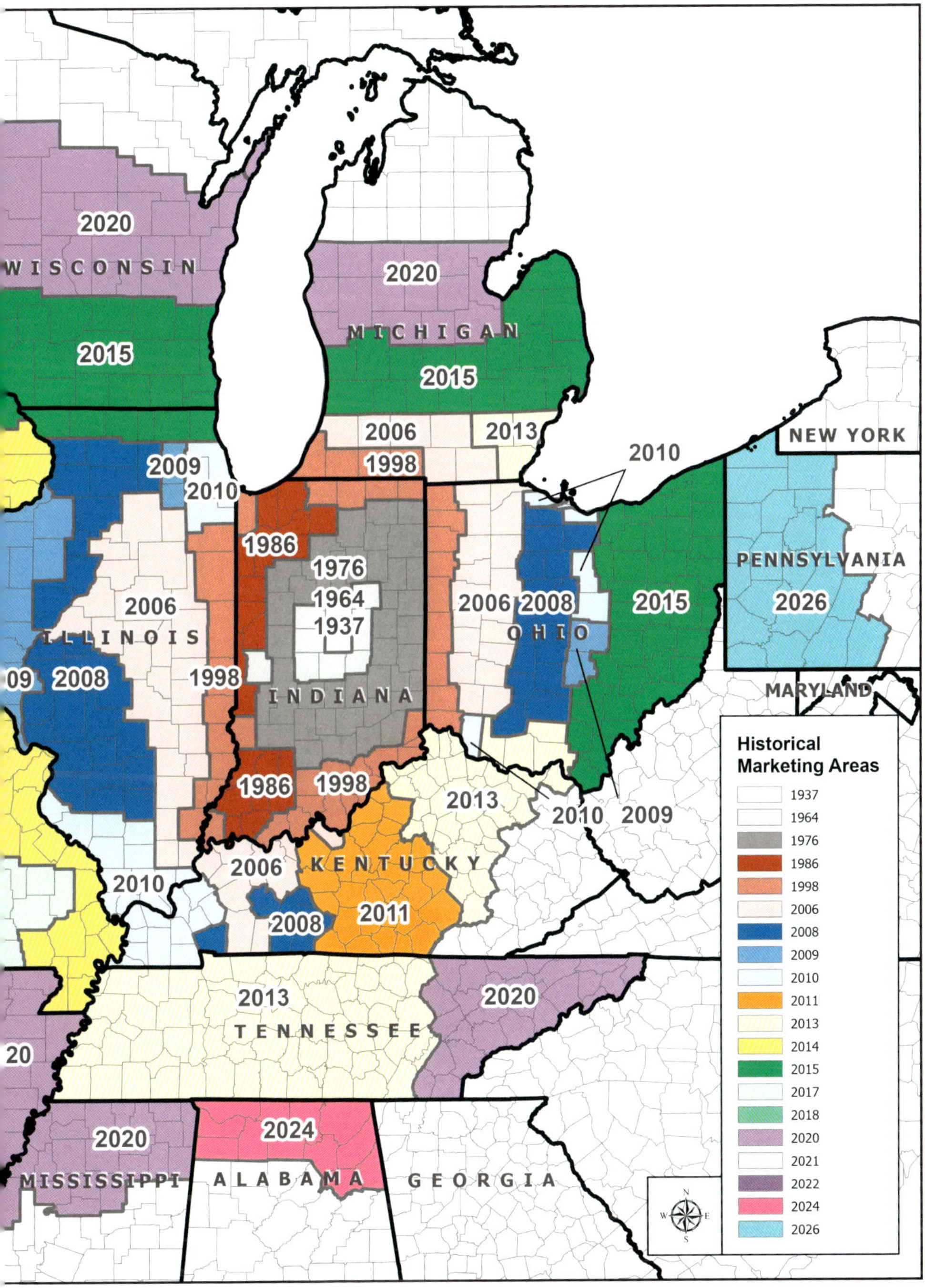
2020
WISCONSIN
2020
MICHIGAN
2015
2015
2006
2013
2009
1998
2010
2010
NEW YORK
1986
1976
1964
1937
2006
2006
2008
2015
PENNSYLVANIA
2026
ILLINOIS
OHIO
09
2008
1998
INDIANA
MARYLAND
1986
1998
2013
2010
2009
2006
KENTUCKY
2010
2008
2011
2013
TENNESSEE
2020
20
2020
2024
MISSISSIPPI
ALABAMA
GEORGIA
Historical Marketing Areas
1937
1964
1976
1986
1998
2006
2008
2009
2010
2011
2013
2014
2015
2017
2018
2020
2021
2022
2024
2026
N
W
E
S

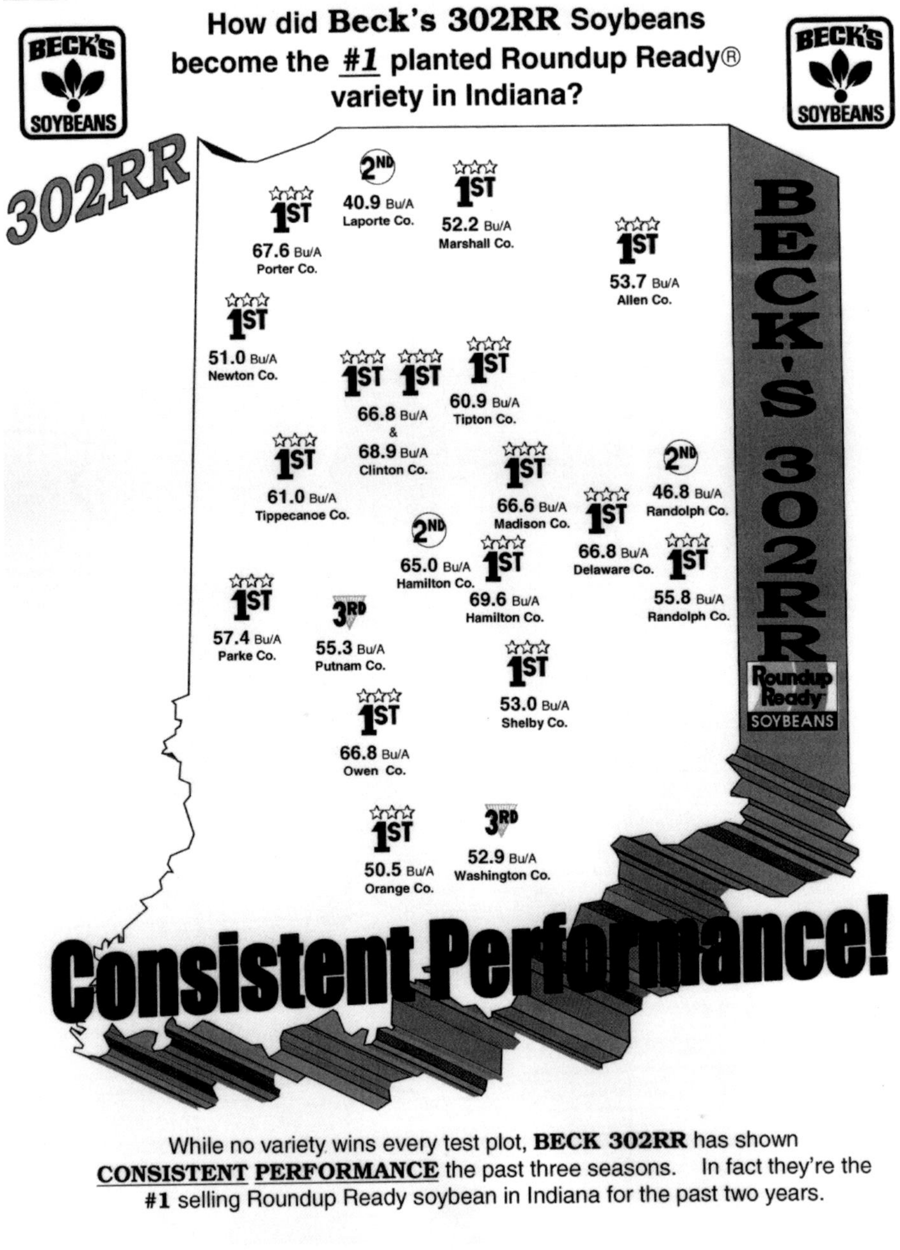

Beck's 302RR Soybean Ad (1998)

Beck's Leadership Team Around the Board Room Table (2026)
From Left to Right: Tom Hooper, Scott Beck, Sonny Beck, Jim Zimmer & Bethany Gremel

Beck's Leadership Team at the Salina, KS,
Grand Opening/Ribbon Cutting at Becknology™ Days (2025)

Sonny Beck and Lee Rulon (1970s)

Lee Rulon and Francis Beck (1970s)

Lee Rulon (1978)

Scott and Sonny Beck

Sonny and Scott Beck in Front of the 300 Bushel PFR Plot (2010)

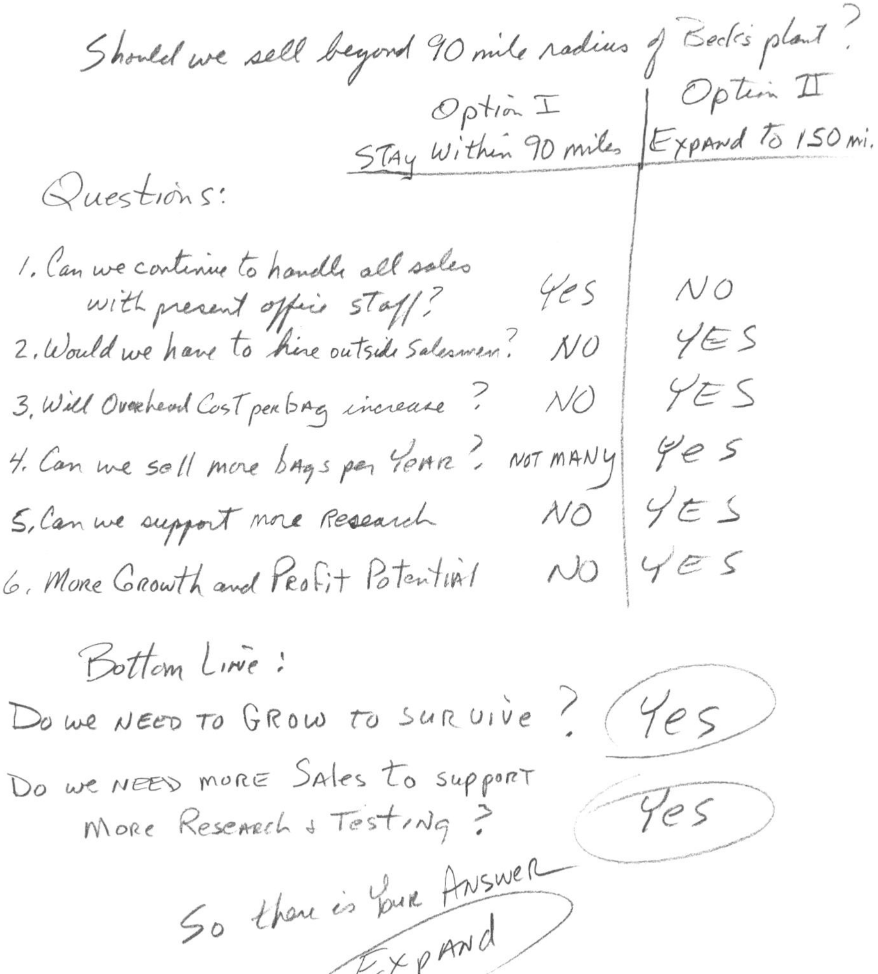

Should we sell beyond 90 mile radius of Beck's plant?

Questions:

	Option I STAY Within 90 miles	Option II Expand to 150 mi.
1. Can we continue to handle all sales with present office staff?	YES	NO
2. Would we have to hire outside Salesmen?	NO	YES
3. Will Overhead Cost per BAG increase?	NO	YES
4. Can we sell more bags per YEAR?	NOT MANY	YES
5. Can we support more Research	NO	YES
6. More Growth and Profit Potential	NO	YES

Bottom Line:

Do we NEED TO GROW TO SURVIVE? YES

Do we NEED more Sales to support More Research & Testing? YES

So there is Your ANSWER

EXPAND

Sonny's T-Bar Used to Determine if Beck's Should Expand Beyond a 90-Mile Selling Radius (1990)

Beck's Leadership Team (2021)

Aerial Shot of Beck's Headquarters in Atlanta, IN (2025)

Pauline, Francis, Sonny, and Glendia Beck (1990s)

Aerial Shot of Beck's Headquarters in Atlanta, IN (1992)

13

THE WIN-WIN WAY

I've always loved to compete, always have, always will. But somewhere along the way, I learned that real competition isn't about beating the other guy. It's about building something that lasts.

When it comes to negotiations, people assume the goal is to "win" at all costs. They look for someone or something to take advantage of with a single mindset: for one to gain, the other must lose. But that's not how I think. A true victory is when both sides walk away proud of what they have accomplished; a true victory is a win-win.

I've always preferred to shake someone's hand and know that the deal will last, not because I got the better of them, but because we both gained something of value. I want to walk away knowing that I've done everything I could to help the other person feel good about the deal because that kind of trust lasts longer than any agreement could.

That belief, that real winning means helping others win too, was put to the test in a big way starting around the year 2001. Beck's was at a crossroads as the seed industry was changing fast. National companies were pouring millions of dollars into research and genetics, developing traits that were transforming how farmers grew their crops. We realized more than ever that we needed access to genetics and traits, but we couldn't outspend these major companies when it came to research and development.

But I'd been thinking about this problem for years. In the White Paper I'd written not long before, I projected that for a seed company to survive the coming wave of capitalistic consolidation, it would have to make strategic decisions to buy, sell, and trade products and information. So, we started researching our options.

By 2006, we saw a real opportunity. Pioneer had excellent genetics and crop traits that could help our customers be more competitive and grow higher yields. Pioneer, in turn, knew who we were, what we were building, and the value we could bring to their business. They respected our culture, our service, and the way we built genuine relationships with farmers.

I still remember the feeling of walking into those meetings, possibility and tension hanging in the air. They had what we needed, and we had a customer base and way of doing business that they respected. But could we find a way to blend the two without losing what made each of us unique?

Negotiations went back and forth; after all, win-win outcomes don't always take the easy road. Getting to the point of mutually acceptable terms wasn't easy, and a positive outcome isn't guaranteed. It takes creativity, patience, and a whole lot of honesty. And it takes humility to see things from the other side. In any negotiation, I always ask myself, "Are we helping them win, too?" Because if we are, we'll win bigger in the long run.

That's the part people don't always understand. A win-win deal requires vision. You've got to see beyond what's on paper and understand what the partnership could become.

Eventually, with a lot of hard work and goodwill from both sides, we found our common ground; our very own "win-win." The result was something brand new: Beck's name on the front of the bag, and the Pioneer "XL" branding right beside it.

The first time I held that bag in my hands, I felt a sense of pride. Not just because we'd secured genetics that would help our farmers, but because we'd done it our way, the right way… the win-win way.

That might sound conceited or businesslike, but for me, it wasn't about economics. It was about people. You can't buy, sell, or trade anything without someone on the other side. Every deal is a relationship, and relationships, done right, are built on trust. You can't build something lasting by tearing someone else down.

Looking back, there are a few lessons that have stuck with me from that experience and others like it.

First, I've learned that the end result must be advantageous for everyone. Another truth I've realized over the years is that when reasonable people share the same information, they rarely disagree. That's why I believe in putting everything out on the table. No secrets, no games. If everyone's looking at the same facts, you can reach a real understanding.

Third, it's important to pay attention to people, not just the numbers. Sometimes there's one person on the other side who's struggling with the deal. Maybe they're worried about how it'll look, or whether they'll have to defend it to their boss. That's the person I focus on. I try to understand their perspective and ask myself, "How can I help them feel good about this?" Because once they do, everything else starts to click.

The final thing I've learned is that you have to always remember what you stand for. At Beck's, that's helping farmers succeed. We can't do that by beating people. We accomplish that by building trust among our partners.

At the end of the day, protecting Beck's means protecting who we are. And when values lead the way, your company will follow.

I saw this same principle play out around 1999, during the dot-com boom. I had a good friend of 20 years who teamed up with two partners and made a go at starting an online seed company. They were talented salesmen and I respected their drive, even if it meant new competition for Beck's.

Not long after they got started, my friend called me for a favor. They were struggling to secure germplasm to put into third-party yield trials; and without it, the company would fail.

He asked me if Beck's would consider supplying them with seed they needed to initiate their hybrid seed corn line. I absolutely understood where they were coming from, and while I could have said "Sorry, you're a competitor of ours, good luck," I chose another path. The win-win path. Instead, I told them, "Of course, we'll help. Glad to do it."

That's win-win, too. Sometimes it's not about contracts or deals or numbers, it's about being there when someone needs you, even if there's nothing in it for you.

When you strip away all the business talk and strategy, a win-win approach really comes down to one thing: relationships. Do your best for people, and they'll do the same for you. That's not just good business, it's good living.

Because when you help others succeed, you've already won.

14

THE HEARTBEAT OF BECK'S

The best measure of a company isn't in sales figures or awards. It's in the quiet moments when values show up in action.

I'll never forget when a dealer told me about his experience attending one of our annual Becknology Days events. He went there almost on a mission to see if the employees he met lived out the culture we talk about so often. He looked closely, watching for arrogance, indifference, or impatience. But what he found was the opposite: smiles, handshakes, and genuine conversations. Later, he told me, "I couldn't find one person who wasn't glad to be there, glad to see me."

That story left me beaming with pride. It reminded me of a simple truth: we're not just a seed company. We're not even just a marketing company. We're a relationship company. And relationships are both the foundation and the pinnacle of what we do.

Culture is the hardest thing to measure, and yet it's the most important thing we have.

At Beck's, our culture isn't something that can be easily defined. It's something you need to experience to truly understand, and we work every single day to protect it. We don't aim to preserve our culture; we seek to improve it.

I often say: if you take care of people, they'll take care of you. That philosophy isn't complicated, but living it day after day takes relentless effort. And it starts with trust.

I've always treated people from a foundation of trust, but it must be earned, not assumed. It's earned in the follow-through, in whether someone does what they said they'd do. Big talk doesn't matter. Action does. Even imperfect action builds trust, because it shows sincerity. Without that, you can't serve others.

My view of culture is tied to my faith. My Christian beliefs guide me to respect others, to be honest, and to treat people the way I'd want to be treated. I always try to put the golden rule into action, doing my best to live it out every day and letting my actions speak for themselves.

As we've grown, from six employees to fifty, to now over a thousand, we've realized that our culture couldn't be left to chance. People don't just absorb it; it has to be made tangible.

In 2010, Scott and I asked a group of managers and directors to describe what made Beck's unique; what sets us apart in what we stand for and how we hire. Some wrote just a few pages, others wrote several. We worked those responses down to one page with six essential truths. Those six points became our compass, and today we continue to hire and live by our "Attitudes and Actions." Many companies have value statements that are often vague and ignored. But ours are core values, specific behaviors, and talents we care about most.

Teamwork. Integrity. Innovation. Adaptability. Commitment. Passion.

But values alone aren't enough; we needed a structure that showed how they all fit together in daily life.

In 2018, still on the quest to summarize our culture in a tangible way, Scott created the "A-Frame" for our business, which illustrates the framework we use to achieve our purpose: Helping Farmers Succeed. At the base is faith expressed through service; namely, honor God and help others. Resting on that base are our attitudes and actions.

That's the house we live in.

One of the best decisions we've made was bringing in Bethany Gremel. Today, she serves as our director of culture and brand experience and a member of our Leadership Team, but she has played a few pivotal roles in shaping our company and our culture over the years. Bethany's roots run deep: her family goes back seven generations as farmers and dealers with Beck's. She left a Fortune 500 company to come home, to carry forward a heritage and traditions that shaped her life.

One of her traditions is to invite every new employee into her home for breakfast. Imagine a member of the Leadership Team standing over a stove, making biscuits and gravy, and greeting our new employees by name. The smell of sausage and coffee fills the air as nervous new hires laugh and begin to feel like they belong. That's not a corporate program. That's culture in action.

And she's not alone. Leaders like Tom Hooper and Jim Zimmer brought in fresh perspectives from big agricultural firms but knew instinctively that culture was our superpower. Together, they have helped frame it into something lasting.

As we have continued to expand into new states and new markets, we've learned the best way to protect our culture is to hire people who enrich it. That means looking at character first and skills second. Out of the 4,000+ applications we receive each year, we evaluate the individual's attitude and work ethic above all else. If you have those, we can probably teach you the rest using the skills you have.

That shift has been transformational. It ensures that as we grow outward, our culture, the heartbeat of Beck's, grows with us.

Culture is built from moments, often unexpected ones that stay with you. I remember an employee whose daughters were graduating from

Purdue. As a Purdue Trustee, I drove up to personally hand them their diplomas. For me, it felt like the right thing to do. For him, it was unforgettable. He later came to me with tears in his eyes, thanking me for honoring his family in that way. That moment reminded me: culture isn't about policies; it's about people.

I've also seen employees grow up here. One young man once told me, "I feel like I grew up at Beck's." He started small, learned, failed, learned again, and over time stepped into real leadership. Watching that kind of growth, not just in skills, but in character, is as rewarding as any business milestone.

If you ask nine out of ten Beck's employees what they love most about working here, they'll tell you it's the culture. It's not abstract. It's alive in trust, in servant leadership, in how we care for one another and our customers.

This isn't about bragging. It's about staying real and helping others. Our culture and our relationships are our light. If we let them fade, the whole place goes dark. That's why we pour everything we have into protecting them and build everything on character and attitude.

It's the greatest expression of teamwork I've ever seen. And it's what will carry Beck's into the future.

15

WHERE APPRECIATION TAKES YOU

Without customers, we wouldn't exist. Sure, ideas and products matter. But the people who believe in us, who plant our seed and trust us with their livelihood, are the cornerstone of our business. My father knew that before I ever did. And he didn't just say thank you; he lived it.

One of my favorite stories about my dad involved one of Beck's very first customers, Paul Wheeler, who was achieving some very unsatisfactory results. Some of the seed he planted came up, and some didn't germinate at all. It didn't take Dad long to realize the farmer had set his new planter too shallow for the seed to access moisture, creating an uneven and unhealthy stand. The mistake was clear, but Dad knew asking the farmer to buy more seed wasn't the right answer.

Instead, he walked alongside him, helped set the planter correctly, and then replaced all his seed free of charge. Then he told him, with absolute conviction, "I'm not selling you a bag of seed. I'm selling you a stand of corn." Those weren't empty words. They were a promise. And it's one we still stand behind today with our 100% Free Replant policy. Trust was the foundation, and appreciation was the way to build it. It's how we grew this business, one handshake and one promise at a time.

In the 1970s, those values took on a new shape with Lee Rulon's direction. To understand why, you have to know the kind of man Lee was. He had a gift with people. Within ten minutes of meeting someone, he knew where they were from, how their crops looked, and what their kids were up to. I admired that because, truth be told, socializing wasn't my strong suit. But I knew enough to trust him.

So, when Lee approached me with the idea of inviting our dealer network to join him on a trip, I knew he was onto something. This trip wasn't just a meeting or a conference, but rather a vacation to somewhere warm and inviting where they could connect, share stories, and learn from one another. He believed that if our dealers could bond with each other, it would strengthen the entire Beck's community. Grand Bahama Island seemed like the perfect destination.

In 1976, Lee packed his bags and headed south with a small group of dealers. I stayed behind because I knew my role. Networking was Lee's gift, not mine, and I trusted him completely. When he returned, the smile on his face said it all. The dealers had laughed, relaxed, swapped farming advice over good food, and left with friendships that lasted far beyond the trip. And it set the foundation for every dealer trip since.

At the same time, Glendia and I were creating our own family traditions. Each year, when the kids finished school and before detasseling began, we carved out a slice of time to get away. In 1986, our daughter Kim graduated from High School and requested we go to Florida in June. We all agreed it was too hot at that time of year. Lee recommended Grand Cayman, a destination he had traveled to before, and we booked our flights for an earlier trip in April over Spring Break.

We stayed at a place called The London House, a new condo complex nestled against a stretch of the most beautiful white sand that, unlike some beaches in Grand Cayman, didn't shift or wash away with the tides. The following year, we just so happened to make a stop in Grand Cayman again on a cruise (Scott's trip request). Sure enough, that same London House caught our eye again. By our 25th wedding anniversary in 1987, when the kids surprised us with a trip back there, it felt like fate. We bought one of those oceanfront condos at The London House.

What started as a family getaway soon became an extension of our business philosophy. We began taking dealers there, then employees.

Before long, it became a reward; an opportunity for hardworking farmers and their families to escape the bitter Midwest cold, sit by the sea, and enjoy themselves without worry.

The trips grew in popularity, and by 2009, we expanded to a cruise. Scott and I each hosted groups of more than 60 customers for a week at sea. Our trips program has continued to evolve, and while the groups grow in size and the destinations change, the gratitude we hear from customers remains unchanged. It's quiet, but it's real, and for me, the reward isn't their thanks. It is the chance to sit back, listen, and learn about their lives in ways I never could in an office.

Over time, we sought new ways to show our genuine appreciation for our customers. We launched the "four-wheel deal," a program where customers who committed to buying a certain amount of seed earned a four-wheeler in return. That simple idea, first imagined by Tom Hooper, has grown into what our Commitment Rewards program is now. At first, it was toolboxes, power tools, and small equipment. As the business grew, so did the value of the rewards.

Today, we don't just see it as a way to say, "thank you," but rather a financial tool that allows farmers to access the equipment they need while freeing up working capital for their operations. With trucks, tractors, trailers, and farm implements available, our customers can leverage their seed commitments to secure significant discounts on the items they need for their operations to succeed. By structuring our program to allow them to commit upfront and benefit immediately, we provide a way for them to enjoy the equipment now while saving money in the long term. It is a win-win, built on trust and genuine appreciation.

Looking back, whether it was Dad handing out replacement seed to a farmer with no questions asked, Lee leading those first dealer trips, or me watching customers enjoy a week of rest and relaxation on a beach in Grand Cayman, the lesson is the same: appreciation is about more than words. It's about actions that prove you care. We don't have investors or outside shareholders. We have customers. And they deserve to feel like more than numbers on a page. They deserve to feel seen, valued, and trusted, just like family.

16

CHASING THE NEW

I stepped into one of our test plots as dew clung to my boots and the soil was soft under my feet. Rows of young corn stretched ahead, each plant a small gamble, each one carrying the promise of what was new. In that quiet moment, it struck me: people in Palo Alto or Austin talk about innovation as though they invented it. But here, in Hamilton County, Indiana, we've been betting our lives on "the new" since 1937.

You might not expect a family-owned seed company, rooted in soil and sweat rather than software, to speak the language of innovation. But the truth is simple: without embracing the new, we'd have disappeared long ago. Instead, we've survived and thrived because "the new" has always been woven into who we are.

For me, the seed has always been the perfect symbol of the new. Every cornstalk that rises from the dirt, every soybean plant, it all begins with something small, fragile, and full of promise. A seed is the very definition of the new.

My dad understood that. In the mid-1940s, he built one of the first known self-propelled detasselers, a machine that removes the pollen-producing tassel from the top of a corn plant to prevent self-pollination and ensure cross-pollination from a specific parent plant. To outsiders, it might have looked like a strange contraption. But for us, it was freedom from one of the most tedious jobs in seed corn production and a leap forward for our business. That machine wasn't just "cool"; it mattered.

I've carried that same mindset into everything I do. I'm a person who believes in improving every year. Sometimes it's as simple as designing a new piece of equipment or building an addition to one of our facilities, and sometimes it's as big as achieving 10% growth over the year before. But you can't get better by standing still. To improve, you must innovate. You must adapt. You must overcome the obstacles. And you must be willing to chase the new.

We integrate that same spirit into everything we do. Every test plot, every inbred line, every cross-pollinated hybrid in our greenhouses highlights our commitment to discovering the new; healthier germination, stronger plants, and higher-yielding product lines.

We have employees whose entire jobs are devoted to the pursuit of the new. Researchers with extensive experience putting them among the best of any agricultural organization in the nation. They'll spend days, weeks, months, and even years examining every tiny detail of a new product.

Our commitment to the new comes from our dedication to the farmer. Every gain the farmer can make with improved seed quality or a new, innovative farming technique means more opportunity for their farm to profit, grow, and survive. It's exciting to show them Practical Farm Research that points to an improvement in their seed performance; it's what keeps them coming back to plant more of our products.

Over the years, we have implemented plenty of new offerings and ideas that have kept us on the cutting edge of our industry. In 1964, fresh out of Purdue and with my dad's permission, I launched our Practical Farm Research (PFR)® program. At the time, it felt bold, even risky. But I believed that farmers deserved unbiased research on products and practices that could help them increase their bottom lines. PFR became the place where curiosity turned into proof.

As universities turned their research dollars into investing in new products, we picked up the slack in providing practical data farmers could use on their farms. We continued to dedicate more acres, employees, and resources to testing factors such as planting date, soil type, nitrogen rates, population, root structure, and tillage practices to help farmers increase their yields without having to invest in the research themselves. To this day, there is no other source of comprehensive, unbiased research on "the new" in the industry.

Other innovations followed. In 1996, we were one of the first seed companies to offer Roundup Ready soybeans. That same year, we introduced SureGro™, our exclusive soybean seed treatment, which became a standard in the industry. In 2001, we changed the game again by treating 100% of our corn hybrids with FaSTart®, a seed-applied insect control that gave young plants a head start. We launched our Commitment Rewards Program in 2009, and then in 2025, Jim Schwartz, director of research, agronomy, and PFR, and his team discovered and patented "Root Reveal." This cutting-edge, remarkable research helps farmers with hybrid selection, product placement, and crop management by uncovering how the root architecture and root volume of each hybrid impact performance. Each of these steps was a bet on the new.

But there's something you need to understand about farmers: they're not dazzled by shiny promises. They respect realism and rely on what they know, what generations before them knew, and what has been proven to work in their operations. I've had farmers look me in the eye and say, "Sonny, don't just tell me this product is better. Show me."

That "show me" mindset runs deep. Farmers know risk. They live at the mercy of weather and markets, but they won't gamble blindly. They want proof; replicated testing, more than one year of trial results, and the plain truth about both advantages and disadvantages. And I respect that, because it keeps us honest. The new only matters when it earns trust in the field.

But the new can't happen without dreamers. The people who dream are the ones who think about doing something different and changing the status quo. I think there's a spark of dreamer in each of us. I try to inspire that in everyone around me. We want every employee, whether they work in the lab, in the processing plant, in the field, in the warehouse, or in shipping, sales, delivery, or accounting, to have the freedom to dream and the confidence in knowing their ideas matter.

Curiosity is part of it, too. It's like dreaming with boots on. Curiosity is the understanding that what we know now isn't enough. It's having the willingness to ask, "What if we tried it differently? What if we risked being wrong in order to discover something better?"

Here's the catch: you can't confine the dreamers to just one corner of the company. The spirit of the new belongs everywhere; the whole organization has to carry it, or it will fizzle out, and the company will grow stale.

None of it works without trust. I've seen teams where the distance between a supervisor and an employee might as well be a hundred miles, even though they sit ten feet apart. Without trust, no one speaks up. With trust, even the quietest voice can bring the best idea forward. Trust creates freedom to dream.

Some of our best ideas come from the people closest to the customer. The quality of the seed, the way a farmer is treated at the farm gate, the first impression someone gets when walking into one of our facilities; it all shapes their experience. That's why we rely on every employee to innovate, not just a select few. Their ideas won't always be perfect, but that's OK. If we wait for perfection, we will never accomplish anything. What matters is progress toward something better, not perfection. Because in farming, in business, and in life, you need the new. And we're not afraid to chase it.

Beck's Seed Bags and XL Brand Bags Featuring the Pioneer "XL" Branding

Sonny, Scott, Corey, and Chris Beck in front of Beck's 400 Bushel PFR Plot Featuring XL® 6622AM*brand Corn That Hit 406.5 Bu./A. (2020)

Sonny Beck in front of Beck's 400 Bushel PFR Plot Featuring
XL® 6622AM*brand Corn That Hit 406.5 Bu./A. (2020)

Sonny and Scott Beck With Production Employees at a Test Plot Featuring Beck's XL Corn

Bethany Gremel Hosting Beck's Employees for a Culture Breakfast in Her Home (2025)

Beck's Culture on Full Display at an Annual Becknology™ Days Event (2011)

2025
BECKNOLOGY
DAYS
BECK'S

Beck's Employees Praying the Morning Before an Annual Becknology™ Days Event (2025)

Scott Beck Presenting Beck's Attitudes and Actions to Employees Outlining What Makes Beck's Unique (2011)

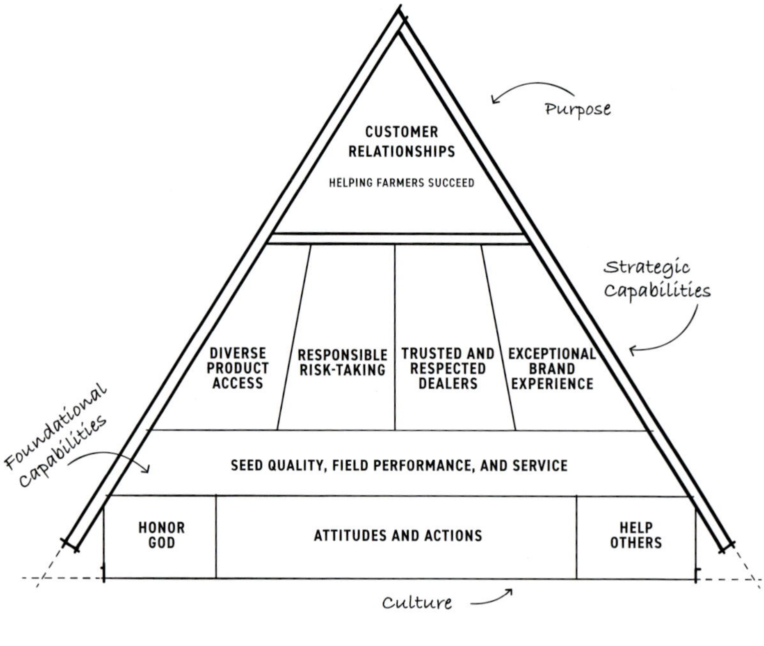

Beck's A-Frame Business Model, Developed in 2018, That Visually Portrays the Company's Foundational Capabilities and Purpose

Todd Marschand, Jim Love, Lee Rulon, Ben Grimme, Kurt Bandel, and Sonny Beck on a Dealer Trip (1993)

Francis and Pauline Beck on a Dealer Trip in the Caribbean

Sonny and Glendia Beck Celebrating Their 25th Wedding Anniversary at The London House in Grand Cayman (1987)

Three Generations of the Beck Family at The London House in Grand Cayman (2012)

As Our Trip Program Continues to Grow and We Welcome More Farm Families to Grand Cayman, Customers Can Now Also Enjoy Staying at The Renaissance, Next Door to The London House

Sonny Beck Speaking at the Grand Opening of Beck's New 1-Million Bag Processing Tower (1998)

Beck's Constructed Their First Two Greenhouses in 2003.
Beck's Now Has Ten Greenhouses at Their Headquarters
Featuring Advanced Climate Controls and Efficient Pink
LED Lighting That Adjusts to Outdoor Conditions.

Beck's Corn Breeding Efforts

Beck Research Employee Conducting Embryo Rescue on a Corn Cob

Beck's Was the First to Market 100% of Seed Corn With a Seed-Applied Insect Control System, FaSTart (2001)

Beck's Introduced Sure Gro, an Exclusive Seed Coating System to the Soybean Market (1996)

Beck's Set Another First By Introducing Escalate®, a Unique Combination of Yield-Enhancing Products That Still To This Day Comes Standard on Every Bag of Corn, Soybeans, Wheat, and Elite Alfalfa (2009)

Sonny, Scott, Corey, Chris, and Jason Beck In Front of Beck's 300 Bushel PFR Plot That Hit 302 Bu./A. (2007)

Scott Beck in Front of a PFR Planting Date Study at Becknology™ Days

Scott Beck Giving the Soybean Product Talk at Becknology™ Days

Every Year, Beck's Shares Our PFR Studies and Data With Farmers at Our Becknology™ Days Events Throughout the Month of August

Sonny Beck Explaining Root Reveal™ Research. Root Reveal Was Discovered by Jim Schwartz and Patented by Beck's. It Utilizes Specialized Root Boxes to Categorize a Hybrid's Root Architecture Into One of Three Structures: Horizontal, Balanced, or Vertical, to Help Farmers Implement Appropriate Management to Maximize Yields.

Jim Schwartz, Director of Research, Agronomy, and PFR, and Dr. Steve Becker, Phenomics Manager, Unboxing a Balanced Root Structure Root Reveal™ Research Plant

17

GLASS DOOR, GLASS FLOOR

The sun was blinding through the pane of glass behind us as I stood next to Indiana Governor Eric Holcomb and lifted the oversized scissors, our fleet of planes towering in the background. I remember feeling proud and a little giddy, like a young boy on Christmas morning. As the ribbon fluttered to the ground, I looked up at the crowd gathered before us as sunlight poured straight through 120 feet of glass, flooding the space with light.

It was a moment I will never forget.

The door, 28 feet tall and 40,000 pounds of solid glass, was something to behold. And while it was just a door for most of the people there that day (albeit a beautiful, massive door), to me, it was a reflection. It reflected everything we'd worked for and everything we stood for: transparency, vision, and the belief that even ordinary things could speak volumes if you dared to see them differently.

In August 2015, we broke ground on our private hangar in Hamilton County near Zionsville, Indiana. It was a big deal for us at that time because the 19,000-square-foot facility was the first private hangar at Indianapolis Executive Airport. More importantly, though, the new facility was crucial for us to continue engaging with new potential customers. For years, we've flown farmers to our headquarters to see not just the product they are buying but the company behind it, and this new hangar would make it possible for us to continue showing prospective customers from hundreds of miles away Beck's culture and experience.

I remember standing over the boardroom table, reviewing the hangar site plans with our facilities team. It was impressive to say the least, and they had incorporated everything we had asked for. But I kept coming back to the retractable door and thinking to myself, "What if this door could say something?" What if the very first thing people saw upon arrival told them something about who we are and our commitment to innovation, transparency, quality, and progress?

What if it were made of glass? I figured the cost would be sky-high, but I asked the team to price it out anyway. To my surprise, it wasn't much more than the standard material, and suddenly, the impossible didn't seem so impossible after all.

Before long, we were designing something no one else had ever done before: a 120-foot-wide, forty-thousand-pound glass retractable door. The door's message is clear; it shows we're big enough to get the job done and bold enough to do it differently. It's a reflection of how we think, how we care, and what we believe, and its power has gone beyond anything I ever predicted. One thing is for certain: every person who steps foot inside that hangar leaves knowing who we are and what we stand for.

We partnered with a company called Woolpert to handle engineering and construction, and my wife, Glendia, poured her heart into the interior design. She transformed the space into something that felt less like a hangar and more like a second home, complete with a welcoming lobby, offices, sleeping quarters, a workout space, and a kitchen. When it was done, we had something one-of-a-kind that truly made a statement and reflected our family and our business. That door set the tone for how we welcome people, how we serve them, and how we think about the future.

A few years later, as we began designing our new, state-of-the-art soybean processing tower, that same mindset came back to me. The building itself was already extraordinary. At 99 feet tall, it is one of the largest soybean processing towers in the U.S., capable of handling four to six million bushels per year. From a distance, it gleams white against the Indiana sky, a symbol of progress and precision.

I couldn't help thinking there might be an opportunity to make a stronger statement inside. Could we deliver the same message that the hangar door delivered into this space, too?

During one of our design meetings, someone suggested adding a balcony so visitors could look out over the operation below. It was a good idea. Then someone else said, "What if the balcony had a glass floor?"

A glass floor. Suspended in midair, allowing you to watch the seed bagging process from above. Like the Skydeck at the Willis Tower in Chicago, a place where you can see the world in a whole new way. The moment I heard it, I knew - that was it. It was bold, but it was perfect.

By the summer of 2021, the tower stood complete with a one-of-a-kind glass floor. But it isn't just impressive; it's symbolic. It says we have nothing to hide. That we stand for clarity, honesty, and vision.

A family-owned seed company with airplanes and a hangar door made of glass. A ten-story production tower with a floor made of glass.

The door welcomes people in.

The floor lets them see what we stand on.

Two ordinary things transformed into reflections of who we are. That tells the world we're not afraid to think differently, to build boldly, to let the light in. From the door to the floor, our message is clear: we're a family that believes in innovation with integrity, and we'll never stop finding ways to make the ordinary, extraordinary.

And every time the sunlight hits the glass, I'm reminded of the same thing: the future is brightest when you build it to let the light through.

18

LEADERSHIP IN MOTION

I never set out to be a leader. It wasn't something I dreamed about or a role I chased down with ambition. I just threw myself fully into whatever was in front of me, whether it was school, family, extracurriculars, or farming. When I was in it, I was all in. And sometimes, leadership found me.

Growing up, it was basketball. On the court, I wasn't thinking about influence or leadership; I was just trying to out-hustle Ron McGill, my teammate and fiercest competitor. Ron and I pushed each other every practice, every sprint, every game. If I stayed late, he stayed late. If he dove for a loose ball, I wouldn't be far behind. Without realizing it, the rest of the team started feeding off that energy. We didn't do it for the attention. We wanted the team to be the absolute best it could be, so we knew we had first to push ourselves to be the best we could be. Inadvertently, the rest of the team followed suit. By my senior year, Coach Leeman handed me the captain's role.

That's when I learned something important: leadership isn't about a title; it's about the effort you put in that others can see and respect. A title may announce authority, but respect earns followership. You can slap the word "captain" on a jersey, but unless people trust your heart, it won't mean much. I've seen both sides, real leaders without titles, and the titled leaders who couldn't inspire a soul.

I think about some of the great Indianapolis Colts teams with Peyton Manning at quarterback. Sure, he was the team captain and the undisputed leader. But many other players like Marvin Harrison and Jeff Saturday were leaders, too. Manning would be the first one to tell you about the impact they had on and off the field.

When I left the basketball court behind for college life at Purdue, I stumbled upon another lesson. In my first two years, I was in the band. Long rehearsals and endless practices left little time for organizations or a social life. Meanwhile, I watched fraternity presidents and sorority leaders rise into roles through endless hours of involvement and relationship-building. When my two-year band commitment ended, I was able to devote more time and attention to the clubs that had always interested me: agricultural groups, FarmHouse fraternity, my church, the Interfraternity Council, Iron Key (a secret, honorary society), and the Student Union Leadership.

I didn't run for president right away. Instead, I spent time building relationships and volunteering. If there was a fundraiser to organize, I signed up. If the group needed someone to call every alum on the list, I picked up the phone. Every task became a commitment I intended to honor, just like I had in the classroom or on the basketball court. And people noticed. They saw two things in me: I always gathered an extensive amount of information before making decisions, and I followed through once I committed.

Before long, I was being named vice president or president of many of the organizations, badges I wore proudly, as it meant people trusted me to make things happen. The title didn't matter as much as the consistency of action. Leadership is simple in theory, but hard in practice: make good decisions and follow through. Information takes time to gather and patience to understand. Following through requires discipline and honesty. Those two things, patience and consistency, were the recipe for my success.

Decades later, Lee Rulon reminded me of that recipe. In the early 1990s, he sat me down and said, "Sonny, you need to step away from the fields and the office more and start building deeper relationships within the seed industry. Start with ASTA (American Seed Trade Association)."

I trusted his input, as he always gave sound advice, so I dove headfirst. Just like at Purdue, I jumped into committees, took on projects, and delivered. I read everything, asked questions, listened, and followed through. And before long, the members voted me president of ASTA. Different stage, same approach.

These experiences outside of Beck's gave me perspective. When you're at the top of any organization, whether a high school basketball team, a college fraternity, or an industry association, people look to you for the final call. You can't hand that responsibility off. You can seek input, you can lean on others, but ultimately, the decision rests in your hands. That weight is real, and it's not always comfortable. But it's the job.

I've also learned that you can't lose your temper. People are always watching, and they'll remember how you act under pressure more than they'll remember what you say in a speech. Blow up once, and trust evaporates. Instead, I've tried to remain steady, keep my decisions clear, and maintain the trust others have in me.

Listening is another cornerstone of leadership. Not just hearing but really listening. People need to know their input matters, that their ideas aren't vanishing into thin air. And sometimes what you hear in one moment proves invaluable in another, years down the road. A good leader remembers those voices and gives credit where it's due. Forgetting to share credit (or worse, taking it for yourself) is the fastest way to lose respect.

I've seen firsthand the danger of leaders who shut their ears, convinced that their success is self-made and self-sustaining. True leadership thrives on listening. It builds trust, it strengthens decisions, and it gives people ownership in the direction you're heading.

So, what have I carried with me all these years? The lessons are simple but never easy: make sound decisions, follow through, listen. Titles may come and go, but respect is built one choice at a time. And in every season of my life, that's the kind of leadership I've tried to exhibit.

19

FINDING MY VOICE

When people ask me about leadership, the subject of communication nearly always comes up. They want to know how you find the courage to stand in front of people, how you learn to share your vision in a way that others want to follow. For years, I thought communication meant speeches; standing stiff behind a podium, delivering lines that sounded polished but often landed flat. I hadn't yet realized that communication wasn't about words, it was about connection.

I wasn't born with the gift of easy conversation. My father and grandfather were much the same: quiet men who let their work and their numbers do most of the talking. But my mom? She could walk into a room and leave with three new best friends. My wife, Glendia, has that same gift. People gravitate to her and feel at ease in her presence. For me, it didn't come naturally. I had to work at it. I had to decide, deliberately, to become something different than what I was.

That decision came gradually, but it crystallized in the early 1990s. I had always been more comfortable with others doing the public speaking, but I began to realize something important: if I wanted Beck's to grow, if I wanted us to be more than just another seed company, I had to grow first. I firmly believe companies don't grow; people do. And if I didn't evolve, the company wouldn't either.

I often think about the idea of two doors. One door allows you to stay exactly as you are; comfortable, familiar, unchanged. The other requires growth. It demands that you become something more than you've been. I knew which door I had to walk through. Choosing it meant forcing myself into situations where I would once have stood back.

At first, it was uncomfortable. But over time, something shifted. I stopped seeing communication as a task and started seeing it as a way to connect farmer to farmer, employee to leader, vision to reality. By leaning into discomfort, I discovered that being more social could actually give me energy, just as it did for my mom and my wife. That shift changed me, and in turn, it changed the trajectory of our company.

I knew that leadership isn't built behind a desk, it's built by showing up. I pushed myself to be more visible, to represent not just Beck's, but our industry. That push forced me to sharpen my communication skills and learn how to carry our message with clarity and conviction.

Speechmaking never came naturally to me. I don't necessarily enjoy writing them, and I've never enjoyed reading from a script or hiding behind a podium. But I developed a discipline that worked. I stripped my talks down to a handful of key points, memorized them, and focused on the stories behind them; the real experiences that connect with people. Because the truth is, nobody remembers a speech word-for-word. They remember how it made them feel. They remember the story.

I learned that the best communicators aren't always the most eloquent-speakers. They're the ones who earn trust. They look you in the eye. They listen as much as they speak. They make you feel like their story is your story, too. And when people trust you, they'll follow you, not because of a title, but because they believe in who you are.

That realization transformed my leadership, our culture at Beck's, and the way we achieve our singular mission of helping farmers succeed. We became a relationship-based company, not just a seed company. In the early days, when I was in the field selling directly to farmers, I learned quickly that data and product information weren't always enough. Some farmers wanted to talk about soil types and yields. Others wanted to talk about their families and the basketball game the night before. Communication isn't one-size-fits-all; it is relational. It means meeting people where they are, building their trust first, and then letting everything else follow.

COVID brought that lesson into sharp focus. The months of isolation were hard for me because I'd grown to rely on those in-person connections. I missed the energy of walking into a room, shaking hands, and looking farmers in the eye and asking, "How are you doing today?" Those conversations remind me that leadership is about relationships, and relationships are built through communication.

When I look back now, I can see the journey clearly. I started out more like my dad; reserved, cautious, and content to let others do the talking. But to lead, I had to grow into something new. I had to step through that second door. And in doing so, I became more like my mom and my wife, the ones who could light up a room simply by being present. If you ask people around Beck's today, they'll tell you I'm relational, that I connect easily, and that I thrive in conversation. And while that wasn't always true, I've learned that change is possible and necessary.

My story is proof that leadership is not static. It is not about who you were yesterday; it's about who you choose to become today. My journey from quiet observer to communicator is a reminder that a company's growth depends first on the growth of its people.

That's what communication has taught me. It's not about perfection or polished speeches. It's about choosing to evolve, connect, and be real. Because companies don't grow, people do. And when leaders step through that second door and find their voice, they bring everyone else with them.

20

THE LIGHT BESIDE ME

I was seventeen, seated in a cabin with a dozen other guys, mostly sophomores in college. We were tucked away on a wooden trail at Camp Minawonka in Wisconsin. One of the leaders asked us to think about the kind of person we wanted to marry. "Make a list," he said. "Ten qualities you want in a wife. And rank them. You won't get everything, so decide what is most important to you."

The air was filled with the smell of pine needles and damp earth. I took off towards the trail and walked with my head down, turning over in my mind the kind of partner I might someday have. It was the first time I had ever considered marriage or what I thought was important for a life-long commitment. I knew I wanted someone who believed as I believed, who shared values at the core, not just hobbies or interests.

I came home from that trip with clarity. I didn't know it then, but almost every quality I listed out described the woman I would eventually marry: Glendia Mosbaugh.

I knew we were about to go off to different colleges, so despite how I felt about her, we agreed it would be best to spread our wings if we needed to. We ended up going steady over the years (for the most part) and she enrolled at Purdue my senior year and even worked in the office of two of my professors while I attended grad school. As fate would have it, we were married on August 12, 1962. And years later, I can say without hesitation that she has been all of that and more. We love each other and respect the different qualities each of us brings to our marriage.

From the beginning, she was not just my wife. She was my partner; in raising children, in carrying the weight of family through every stage of life, and in our business where she worked as the secretary, typing and printing the test plot books, and preparing lunch every day for those of us in the office. She is good at the things I am not, and she is everything to me.

When we first started dating, we found joy in the easy things: we both loved sports and just spending time together. But I quickly learned that real connections aren't built on trivial things. What mattered most was that we believed in the same foundations: faith, family, and hard work. If those values hadn't aligned before we said, "I do," I'm not sure where we'd be today. You can survive different preferences in sports teams; you can't survive different foundations for living.

One blessing I took for granted back then was how our families embraced each other. Our moms were friends before we were even born, so we knew each other our whole lives. My parents loved her, and hers loved me. Looking back, I realize how rare and important that was. Marriage doesn't just join two people; it ties two families together.

I wasn't always the most social person. Talking to strangers didn't come naturally to me. But with Glendia, conversation was effortless. From our earliest days, we could talk about anything: faith, fears, hopes, disappointments. More than sixty years later, that's still true.

We have disagreements, those times when we see things differently. She'll tell you that I'm super-competitive and analytical. I know that can drive her crazy sometimes. We just don't let those times of disagreement hang on and become something bigger or longer lasting; we talk about it. We know each other, what we value, and what we base our lives on. And in the end when reasonable people have the same information, seldom do they disagree to the point where they are unable to reach a consensus.

I'll never forget a conversation we had in the early 1990s, when the seed industry was shifting, consolidating, and changing faster than any of us could predict. I had been working on the White Paper, laying out a roadmap for how Beck's could survive and thrive in that new world.

I was already working long hours. And Glendia had been watching me wrestle with the weight of it all. She knew that this new roadmap would require me to give even more of myself, my time, and my energy to the company, but she also knew it was something I had to do.

We came to an agreement. I could take another 10 or 15 hours each week over the next few years and dedicate them to the business. Our kids were grown, but she would be there to help with the grandkids, our parents, and whatever came up at home. She told me to do what I needed to do to make the company what we both knew it could be. And after five or ten years, we would reassess.

That was no small offer. She knew it meant nights when I'd stay at the office long after everyone else had gone home. She knew it meant that if something came up at home, unless it was truly serious, she'd shoulder it. And she never once used that sacrifice against me. Yet, there were times when she felt alone, and it was hard for her to continue to give up time with me. But she continued to give of herself, so I could grow the business. That's the kind of woman she is. That's the kind of marriage we have.

Glendia brings so much to who I am and what we've been able to do at Beck's. I've mentioned before that I wasn't naturally a talkative person. Glendia, though, is just a gifted person when it comes to making people feel at home, comfortable, relaxed, and at ease. She's so good at talking with people and having real conversations with them.

Over the years, I've come to see how deeply we complement one another.

While I can be competitive, she has a gift for people, for listening, and making them feel seen, at home, and comfortable in their own skin. I've watched her transform company gatherings, turning acquaintances into friends. The value of her ability to relate to people is so big it can't be measured. Every year that passes, I see it much more clearly.

At Beck's, while I was thinking about the future and strategy, she was making sure people felt valued and connected. Both roles mattered. Both roles helped build the company into what it is today.

I think about how different Glendia and I were socially, and it's a great example of the way a marriage has to work. Marriage is about completing one another, not competing against each other. She's the one with the eye for creating a warm home; I'm content just being there. She thrives in social gatherings; I follow her lead. Together, we form a team where each plays a role, and together, we win.

I realize now that the foundation of our marriage was laid long before we said our vows. It was there through shared values, supportive families, honest conversations, and an unshakable belief in each other. Still today, I can say this without hesitation: every good thing in my life has been multiplied by Glendia. She is my partner, my encourager, my challenger, my balance. And she has been the steady light not only in my life, but in the lives of our children, grandchildren, and now great-grandchildren.

One of the greatest lessons I've learned is this: leadership begins at home. Choosing a partner who believes in your purpose and is willing to share in its sacrifices shapes every part of your life. The right person doesn't hold you back; they help you move forward with confidence. That's true in marriage, in business, and in life. The people beside you matter more than most decisions you'll ever make.

Beck's Fleet of Planes at The Private Airplane Hangar in Zionesville, IN,
Featuring a 120 ft. Wide, 40,000 lb. Glass Door

Sonny and Scott Beck With Governor Eric Holcomb and Other Employees and County Officials at the Ribbon Cutting of Beck's Private Airplane Hangar (2016)

Sonny Beck Delivering a Speech at the Grand Opening of Beck's Private Hangar (2016)

Sonny Beck, Beck Family Members, and Beck Leadership With Governor Eric Holcomb and County Officials at the Ribbon Cutting of Beck's East Soybean Tower (2021)

Bottom View of the Glass Viewing Deck in Beck's East Soybean Tower (2021)

Top View of the Glass Viewing Deck in Beck's East Soybean Tower (2021)

Feb 1993 1

How are we doing? At BECKS

State of company message, FOR 1992

A. Sales were up corn & Soybeans & wheat.

B. Expenses were up

C. Net about same as 1991. We are going to be one of the survivors!

D. FOCUS FOR 1993

1. Being effecient – If not busy think of something to do or ask.

2. Guys in production & Keep Sonny informed.

Sonny at more & more meetings

Need to give me latest updates

Almost need production manager. Another cost Another level And I like to get info 1st hand.

It will work if you report. things that effect next days planning.

3. FOCUS on Customer

A. Do what it takes to give him good service.

B. Smile, be friendly, Don't let them know you're having a bad day.

That includes all of us.

So for 1993 FOCUS on Being Effecient & Serving the Customer

E. things to watch out for in 1993

1. Production – Be careful Be safe Don't get hurt.

2. SAles – Don't be surprised if Pioneer picks on us. They have done it to other small companies in other states. I mean Special deals to Beck customers etc.

But We will survive. As a team we'll survive.

Sonny Beck's Handwritten Notes When Crafting His Annual State of the Company Message to Beck's Employees (1993)

Sonny Beck Leading a Meeting Where He Explains Beck's Access to Genetic Diversity and Technology Through Relationships With Other Companies

Sonny Beck Presenting at an Industry-Hosted Event

Sonny Beck Presenting the President's Message at an Annual Becknology™ Days (2010)

Sonny and Glendia Beck at the State of the Company Message Where Sonny Was Recognized for Winning The Sagamore of the Wabash Award (2024)

Sonny Beck Sitting Down to Film an Episode of Across the Acres - a Beck's Podcast (2025)

Sonny Delivering a Speech at the NAMA Agri-Marketing Conference (2014)

Glendia and Sonny Beck in Front of Their Residence That Has Been Home to Four Generations of Becks

Glendia and Sonny Beck Freshman Year at a Purdue Valentine's Day Dance

Sonny and Glendia Beck at a FarmHouse Fraternity Dance at Purdue

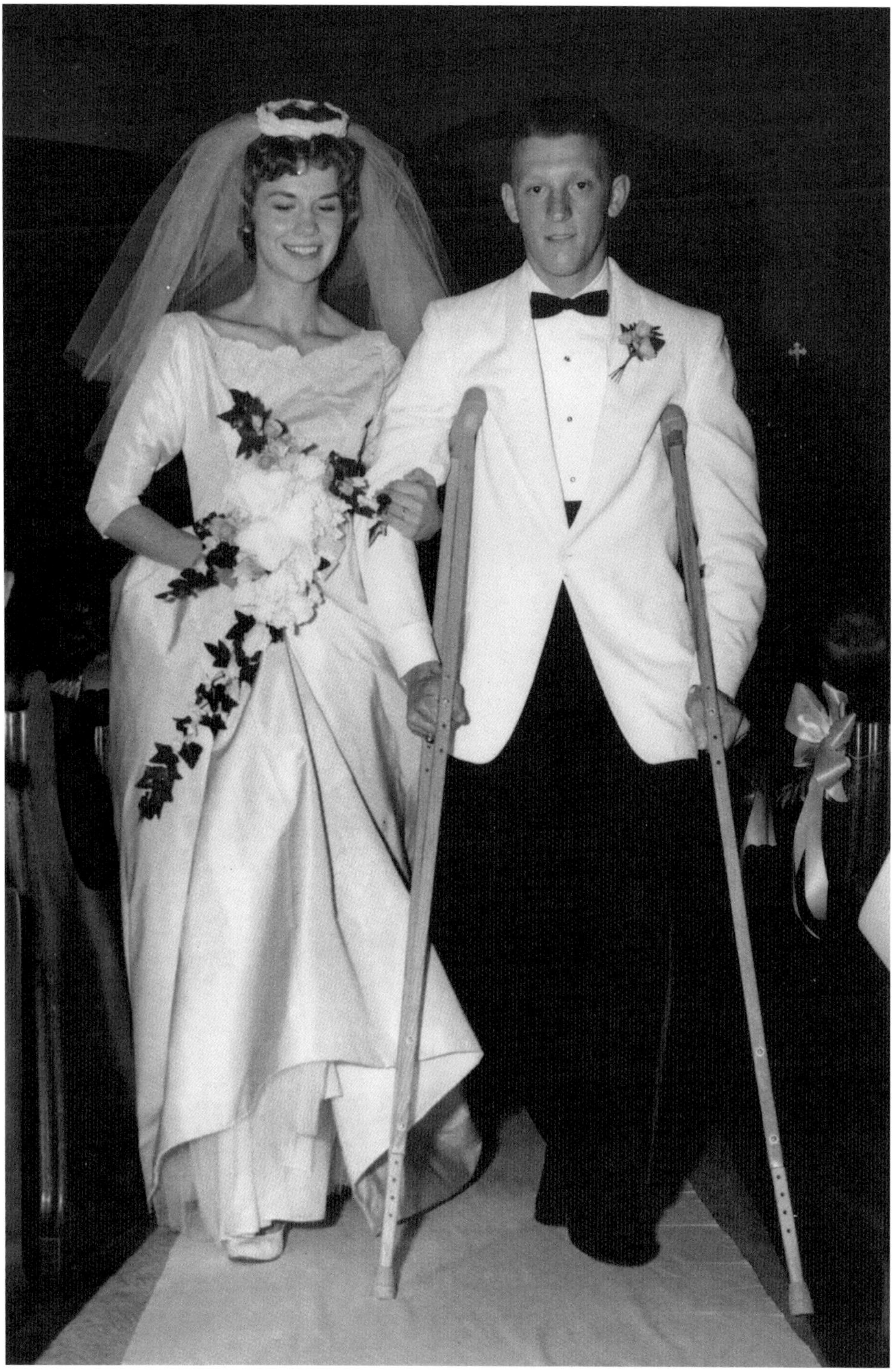

Glendia and Sonny Beck (With a Broken Leg) Wedding Photo (August 12, 1962)

Sonny Beck, With a Broken Leg, Sitting Back Watching Glendia Change a Flat Tire En Route to Their Honeymoon (1962)

Glendia Beck Changing a Flat Tire En Route to Their Honeymoon (1962)

Glendia Beck Answering the Phone at Beck's Office

Glendia Beck Working at Beck's Office

Sonny and Glendia Beck in the 1970s

Sonny and Glendia Beck

Sonny and Glendia Beck on a Dealer Trip in Grand Cayman (1990s)

Glendia Beck at Becknology™ Days

Sonny and Glendia Beck in Their Backyard

Sonny and Glendia Beck Speaking on a Family Panel During Becknology™ Days

Glendia Beck Sitting Down for an Interview for Beck's Home-Grown Legacy Documentary (2024)

21

THE FATHER I BECAME

The first words I ever heard as a new father weren't profound; they were simple: "You can come in now."

Back then, dads didn't pace beside hospital beds or hold their wives' hands through delivery. We waited, awkwardly and anxiously, until a nurse poked her head through a door and waved us in. That's how it was when Scott arrived in 1965, again when Kim came along in '68, and later with Tony in '72.

No matter how many times it happened, walking into that room always felt like stepping into a brand-new world. One where I was suddenly responsible for something more important than land, seed, or even the family name.

We started our family about two years after getting married. Looking back, it amazes me how young we were, wide-eyed, hopeful, and working hard in a world that looked very different from what it does now. We lived in the house my grandfather built in 1912, a sturdy place made from lumber he cut right off the farm. The living room stretched twenty-four feet long, and, in time, that room became the kids' personal jungle gym.

Many evenings, I'd lie flat on my back in the middle of that long room, knees bent, ready. The kids would take a running start from one end, little feet pounding on old floors, and leap toward me. I'd catch them mid-air, flip them over, and send them sailing onto a pile of beanbags six feet away. To this day, when I think of that house, I hear laughter before anything else.

Just sixty feet from that living room sat our seed corn production facility. Back then, it was nothing fancy; just a few wooden bins we had converted into offices. That's where Glendia worked, doing the secretarial duties that kept our small business moving. She could walk back and forth throughout the day, checking on the kids, answering phones, keeping our home and work intertwined in a way that only farm families and working mothers truly understand.

Preschool wasn't really a thing then, not in our world. Most of what our kids learned before school came straight from Glendia. She made sure they learned their letters, numbers, and manners; she was always busy and always involved.

Musical education was another thing we wanted for the kids. My mother was a musician and insisted that I learn an instrument too. I remember telling her I didn't want to carry around anything big, so a trumpet made perfect sense to me. My mother didn't share my logic and came home one day with a trombone, saying, "Sonny, this is what you'll be playing." I was disappointed, but in the end, she was right, and I grew to love it. Likewise, each of our kids found their own musical paths: Scott on drums, Kim on piano, and Tony on alto sax.

Outside of the home, I took the kids with me on the tractor whenever possible. My dad had done the same with me, and there's something about sitting side-by-side on a tractor seat that brings generations together. Farm kids grow up doing things other kids can only imagine. It's one of the greatest privileges of farm life, and I wanted my children to feel it the same way I did.

Sundays were sacred, not just for worship but for us as a family. Farmers usually work Saturdays, so Sunday became the one day that felt like ours. After church, I'd ask the kids, "What do you want to do?"

"Throw the football!" Scott might say.

"Ride dirt bikes!" Tony would chime in when it was his turn to decide.

I'll admit, I loved the dirt bikes. Not just because they were fun, but because they let me take a casual look around the fields and scout to see which ones would be dry enough to work on Monday. They didn't know then how much work and play were intertwined for me. What mattered most was that the kids knew this was our time, it was sacred, and nothing came before it.

All three kids did well in school, and that was a blessing. They didn't need pushing to earn good grades. They were motivated by something inside them, not by pressure from Glendia or me. They played sports too: football, track, and basketball for Scott and Tony, track and cheerleading for Kim.

Tony had my agronomy instincts and thrived when walking fields, studying crops, and understanding the land. When he moved to Illinois to take over the management of one of our production farms, he took my original dirt bike with him so he could scout fields, just as he had with me when he was younger.

Kim was strong, balanced, agile, and quick on her feet. She was always the one lifted to the top of the cheerleading pyramid. Watching her run track or cheer with confidence made me proud in a way that I still remember. Today, she runs our Country Store, connecting our brand with thousands of farmers and their families.

At the heart of parenting, a father must set expectations clearly. My dad was strong in that; he laid out expectations once, and that was enough. "Son, I'm only going to tell you once…" meant exactly that, and I tried to raise my kids the same way. Expectations create atmosphere, and atmosphere shapes behavior. It's not much different from Beck's, where our family of employees know what's expected of them and trust themselves and each other to follow through.

I don't have many memories of punishing the kids, but one significant moment with Scott stands out. He was about three years old, and he darted across the road in front of our house without looking. I saw him and felt a jolt of fear surge through me. I ran, scooped him up, and gave him a swat on the behind; not hard, but firm enough to lift him slightly off the ground. The surprise on his face told me everything I needed to know; he wouldn't do it again. That was the lesson; the expectation was set. There was no yelling or drama, just a clear boundary he'd remember.

A father's example carries more weight than any rulebook. Kids watch you closely. They hear what you say, but they study who you are. Hearing the words, "I want to be like you, Dad," is one of the greatest honors and heaviest responsibilities a man can hear.

I learned over the years that discipline only works if the child buys into it. You can't force someone to hold a value they don't believe in. When my dad once offered me a beer after the hay baling season, I chose not to drink it; not because of anything he said, but because I knew it might affect my sports performance. That self-discipline was mine, and that made all the difference until I was 21.

As a father, the hardest work you do is the work no one sees. It's building the life you want your family to inherit, then living it with as much integrity as you can. If you mess up privately, you pay for it internally. But if you mess up in front of your children? You let them down, and that can be devastating.

Temptations today are stronger than the ones I faced growing up, but the principles are the same: understand the consequences, know yourself, and make choices that align with your values. That's the essence of self-discipline.

When my kids faced disappointment, I tried to do one thing first:

understand what caused it. Was it within their control or outside of it? What could be fixed? The wrong move, in my opinion, is diving headfirst into blame. Fix the problem first, then sort out the "how" and "why." A father should be a helper before he is a judge.

Of all my responsibilities as a father, the most important to me was being present. I often talk about making the most of the 168 hours in every week. If work takes 50 of them and sleep takes another 50, that still leaves 68 hours for your faith, family, and friends. Most of those precious hours belong to family. When you're with them, you must truly be with them.

Turn the phone off. Turn your mind off work. Give your family the best of you, not the leftovers.

A family is a team, just like any team I've worked with at Beck's. You may be the parent, the one with authority, but shared power is better. Respect what each person brings to the table, as success is rarely an individual achievement.

When I think about fatherhood, about the years spent raising Scott, Kim, and Tony, my mind always drifts back to my own parents. They worked the same soil I stand on today. They shaped me through their expectations, their quiet lessons, and the lives they lived right in front of me. The father I became is the man they raised.

Fatherhood helped me understand that being there matters more than doing everything right. I learned that leadership isn't measured only by what you produce, but by how present you are with the people who depend on you. Whether at home or at work, the moments that shape others rarely happen on a schedule. Showing up, listening, and paying attention often matter more than any performance or result.

22

FAITH IN ACTION

Faith has always been one of the three pillars of my life: faith, family, and farming. Without all three, life feels unsteady. Together, they keep me balanced.

But I didn't come to that conclusion sitting in a quiet church pew or by memorizing verses from the Bible. Faith, for me, has always been something lived, not spoken. I learned it from my parents in their everyday actions during breakfast, lunch, and dinner. I learned it in locker rooms with teammates, in fields while driving a tractor cultivating weeds under Indiana skies, and later in boardrooms filled with difficult decisions.

When I think back to childhood, I can still see the devotional books and Bible that sat near our kitchen table. My mom and dad made sure we read from it every evening after supper. We were, of course, in church every Sunday where mom played the organ and dad held leadership roles as an usher and in finance.

What stands out most isn't the Bible itself or the church services, but how their faith showed up in everyday life. I watched Dad treat neighbors and customers with patience. I saw Mom give her time freely to church committees, not because she had to, but because she believed helping others was simply the right thing to do. Faith was never pressed on me with lectures or warnings. It was modeled, quietly and consistently.

When it came to school and sports, I realized something else: life just worked better when I lived out the principles I'd seen at home.

If I treated teammates the way I wanted to be treated, trust grew. If I worked hard and helped others, I felt happier. Simple as that.

As I've gotten older, my conviction has only deepened. Faith is strongest when it's lived, not preached. People notice when your faith looks like service and when your actions match your beliefs. That's why I don't brag about being a Christian. I'd rather people figure it out by watching how I live, how I lead, how I listen, how I serve, and I how I sometimes fail and have to ask for forgiveness. When your actions are genuine, when you give and show up, people can sense what's real. And often, they'll ask the most meaningful question of all: Why?

That's when the conversation isn't forced. It's honest. It's born from curiosity about what lies beneath your choices. And that's when faith has the power to inspire.

Faith-by-action builds on things that people share rather than isolating on things where people differ. I'm not saying differences in faith are trivial or unimportant. They matter, and I'm aware of them. There are all kinds of churches. I'm only saying that I want people at Beck's to feel included and involved in something bigger than themselves; in something that brings out the best in each of us. To me, helping and serving other people is a vital part of it.

I think back to the 1960s and 1970s and realize that most people I encountered probably thought a lot like me. That fact has changed drastically, so I'm more motivated than ever before to live out my faith and let my actions speak the loudest.

Some people wonder how my personal belief in actions-over-words lines up with Beck's public displays of faith. After all, we're proud of our faith and display it everywhere, from trucks and billboards to social media, and even the digital screens in the lobbies at our facilities.

Doesn't that look like talking more than doing?

For me, it comes down to two things.

First, the company has always been an extension of my family, and faith has been part of our family's story from the very beginning. Our daily decisions as a company are rooted in the same principle Jesus taught: "Do unto others as you would have them do unto you." That Golden Rule is the framework for how we work with farmers, dealers, and our partners every single day. Sometimes in meetings, I even find myself reaching back to a Biblical story to make a point, not because I want to preach, but because the wisdom still applies.

Second, farming itself has always carried a spiritual dimension. Farmers live with forces beyond their control. They pray for strength and guidance. They draw resilience from faith. Even scripture is filled with stories drawn straight from the soil: parables of seeds and harvests.

So, when Beck's displays a verse or shares a message publicly, it isn't about boasting. It's about speaking a language farmers understand. It's about saying: We see you. We respect you. We are just like you.

I believe that as a company, our faith is who we are, what sets us apart, and what identifies us. And at the end of the day, my faith is about how I choose to live, and if that can have a positive impact on someone else's faith, then I've done what I set out to do.

23

TIME IN A BUCKET

Family and business have always been woven together in my life, so tightly that it's often hard to tell where one ends and the other begins. Since our founding in 1937, five generations of Becks have carried the torch of our name, each one shaping this company and our way of life in their own way.

The only one I never met was my grandfather, Lawrence Beck. He passed away two years before I was born. Yet even without knowing him, his presence has always been there.

After him came my parents, Francis and Pauline. Then there was Glendia and me. We raised Scott, Kim, and Tony. Then came our grandchildren, Corey, Chris, Jason, Jordyn, Jadyn, Emily, Ryan, Holly, Kiley, Hallie, Hannah, Faith, Laura, and Elsie. And now, I've had the blessing of holding seven great-grandchildren in my arms. That's six generations of Becks who have carried the Beck name, who will play a role in this farm, this company, this way of life.

Whenever we gather together as a family, I am reminded of how lucky I am to be surrounded by so much love. To see generations laughing and talking around the same table, kids chasing each other through the house, and grandkids swapping stories about marriage, life, and work. It's in those moments I am reminded; this is legacy. This is the greatest privilege there is. And seeing three generations actively involved in the business today fills me with pride deeper than words can hold.

And yet, when I reflect on those generations, a sobering fact always comes to mind: less than one percent of family businesses survive to the fifth generation. Less than one percent.

I want us to remain a part of that fraction of a percentage. I'd like to think that Beck's can keep this going, not just as a business, but as a living expression of who we are.

People often ask me how we've managed to keep our family so close and connected while running a growing business. My answer usually starts with something simple: how we use our time.

Over the years, I've had employees come to my office carrying the weight of home troubles or personal struggles. I always listen first. Then, when the moment feels right, I tell them a story I've told a hundred times before. "Every week," I'll say, "you get a bucket with 168 hours in it. No more, no less."

"You can't ask for a bigger bucket," I tell them. "It doesn't matter if you're the CEO or you're just starting out, if your kids are in three sports or if life feels like it's spinning out of control. Everyone gets the same 168 hours. What you can do is decide how to pour it out: work, family, faith, rest. And if you're smart, you'll save a little for yourself."

I believe that a balance in how time is distributed is the key to happiness.

Sometimes they smile, sometimes they nod slowly, and sometimes they just sit there, quiet, letting it sink in. And truth be told, I need that reminder as much as anyone. Some days it feels as though the weeks slip through my fingers in the blink of an eye. But the time in the bucket never changes. It forces you to decide what matters most.

That's been an unwavering truth for me, for Glendia, and for our kids. Pour into the things that matter. Don't waste your hours on what won't last. After all, when the bucket runs empty, you don't get it back.

Growing up, my parents didn't lay down many rules. What they gave us instead was something stronger: expectations. Unspoken, but clear. You knew how you were supposed to act. You knew you didn't want to disappoint your parents or yourself. That invisible guardrail guided me far more than a list of dos and don'ts ever could.

Glendia grew up with the same approach, and together we raised our children the same way. Not many rules, but a shared understanding of what was expected. I believe that shaped them as much as anything we could have said out loud.

I like to think I brought that same spirit into Beck's. We've never been a company with endless rules and restrictions. Instead, we've built a culture of respect and trust where people are counted on to do the right thing because they want to, not because they're afraid of punishment.

When Scott became president, I watched with pride how he carried those same values forward in his own way. He and I are very different people with very different leadership styles, but I like to think we complement each other well. Where I'm wired for vision and competition, he's patient, steady, and thoughtful. Where I push, he listens. Where I charge ahead, he waits for the right moment.

But when it comes to time and family, he's as disciplined as anyone I know. When it's family time, it's family time. Scott and Shantel were intentional about spending time with their kids in many different settings. Kim and Todd and Tony and Tracey all raised their children with their unique approach as well. They never just asked, "What do you want to be?" They instead ask, "What brings you joy?" Over the years, they've helped their kids discover who they are, just as Glendia and I tried to do for them.

And now I see it in our grandchildren, too. Each one carries different strengths. They all have different passions and different ways they find joy, and they all respect one another. I remember watching Bethany Gremel interview Scott's sons, Corey, Chris, and Jason, on stage once. She asked about basketball, and their answers had the whole crowd laughing as they teased each other about who was the better player. But beneath the banter was something deeper. Corey is a thinker and strategist. Chris is all energy and risk-taking. Jason is fiercely competitive and persistent, but being the third boy and the middle child, what would you expect? When I think about all my grandchildren, it's amazing to witness thirteen strong, independent, differing personalities complement each other in life and for some of them, in work.

That word, complement, matters to me. Not "compliment" with an "i," but "complement" with an "e." It's what happens when different pieces can work and blend together into something greater. It's what's allowed Scott and me to lead side by side for years. And it's what gives me hope that the next generation will do the same.

In 2017, our family met to discuss and set requirements for the fifth generation to come into the business full-time. We spent time debating whether they were required to work elsewhere or not. We decided they needed to have at least two jobs working for another boss before they could apply for a full-time job at Beck's. Those two jobs could be internships, part-time, or full-time, but the idea was simple: before they could work here, they needed to see how the world worked out there. They also needed to understand that there are worse bosses out there than their dad and grandpa.

We knew when they returned, they would bring back more than experience. They would bring fresh ideas, broader perspectives, and new energy, adding value to Beck's that we might not have found otherwise.

And once they decide to join the company full-time, we work to expose them to many areas of the business. It was our way of saying:

"Everything here matters, and everyone here matters. Learn it all. Respect it all."

Sometimes I look at families like the Rockefellers, dynasties that grew vast and powerful, and I see how the youngest generations drifted away. Wealth, ease, and distance pulled them from the spirit that built it all. That's why less than one percent of family businesses makes it to a fifth generation.

But I'm an optimist, and I believe we'll beat the odds. I believe we'll see an eighth generation of Becks carrying this work forward, if not more. Because in the end, survival doesn't come down to wealth or strategy. It comes down to how you pour your bucket each week. It comes down to keeping faith, family, and integrity at the center of everything. If we keep doing that, then the seed planted generations ago will keep growing for generations to come.

24

THE FOURTH QUARTER

A friend of mine once told me about the "four quarters of life." He said you can divide a lifetime into 20-year segments. The first quarter (up to age twenty) is largely dependent on others and all about growing up. It's about getting an education and starting to figure out who you are. The second quarter is work and achievement: building a career, starting a family, chasing goals. The third quarter is when you settle into your life's work and carry responsibility for those who depend on you. Then comes the fourth quarter, the years between sixty and eighty, when you think less about yourself and more about what you'll leave behind.

It made sense to me when I first heard it. But looking back now, I can't say my own fourth quarter has gone by that script.

Some folks dislike four-letter words. I despise a six-letter one: retire.

I'll tell you why.

I met Dick Kruse years ago through FarmHouse fraternity. I'd been at Purdue, he'd been at Kansas, and though we never overlapped in college, we both wound up at a special "Seven-on-Seven" alumni event. Seven grads, seven students, one mission: share what we'd learned in life so far.

That was where Dick explained the "four quarters of life" to me. His words resonated, but when I measure my own years against that model, it doesn't quite line up. While many of my high school classmates hung up their hats after thirty years on the job, I was just hitting my stride. For me, everything started coming together at fifty, an age when many people start thinking about winding down. That was the year I wrote the White Paper that helped chart Beck's future. And it was only the beginning. By the time I turned sixty, Glendia was encouraging me to put the foot on the gas and work an extra ten hours a week to get Beck's to where we both knew it could be.

Some people wonder why I never retired. I ask myself a different question: why would I?

At fifty, instead of slowing down, I was learning more than ever: about consolidation in the seed business, about capitalism, about the stewardship our family company needed to survive. I knew what was at stake. If I'd stopped then, I would have missed some of the best years of my life.

I would have missed the teamwork, the thrill of helping farmers succeed, the joy of walking a field at dawn, and knowing we were making a difference. I would have missed the sight of thousands of people pouring into our annual Becknology Days events, families gathering, farmers asking hard questions; everyone sharing a sense of belonging.

If I'd quit, none of that would have been mine to see.

There's a saying: love what you do, and you'll never work a day in your life. For me, that love has always been farming. Planting crops. Building relationships. Helping others grow. That's what keeps me scouting fields and showing up to the office. It's what made "retirement" feel like a foreign word.

Some folks tell me I've been blessed with good health. Maybe so. But I tend to think my health is good because I keep working, not the other way around. The idea of doing nothing and filling days with empty hours has never appealed to me. For me, that would be misery.

Another reason I've stayed so engaged? I've always turned life into a game of self-competition. When I was younger, I'd challenge myself while plowing or planting corn rows: could I make the next one straighter, cleaner, faster than the last? Later, I asked bigger questions: could we improve sales by ten percent this year? Could we help ten percent more farm families? Could we stretch ten percent farther in service, stewardship, and innovation?

That ten percent goal became my way of keeping things moving forward year after year. It's open-ended. It never finishes. And that's the point. If you live by one-off goals, retire at sixty-five, cross a finish line, call it done, you'll stall. But if you keep reaching for more, the years don't matter so much.

But as I continue thriving in my fourth quarter, one thing has become abundantly clear: none of it matters without people. I need the strong, smart, capable people that have surrounded me. I need their energy, their vision, their values. Without them, the work would be hollow and meaningless. But with them, every victory, every milestone means so much more.

That's why whenever I'm handed an award, I don't take it for granted, and I don't receive it on my own. I accept it on behalf of every family member, employee, dealer, and friend who stood shoulder-to-shoulder with me. Because the truth is, we've earned it together.

If I had to boil it all down, the lessons I've learned in my fourth quarter all come back to mindset and attitude. The only thing that matters is the outlook you bring to every challenge you face.

That's how I see my fourth quarter. Not a time to rest, but a time to charge, head high, heart full, with the people beside me who make it all possible. And that's why I don't believe "retire" will ever be a word you hear from me.

EPILOGUE

The sun was setting on a breezy spring evening in 1973 when I pulled the planter out of the last field. Dust hung in the air behind me, and I couldn't wait to get home to the supper Glendia no doubt had waiting on the table.

Off to the side of the road, I saw a familiar old truck creeping along. It was our neighbor. He slowed, rolled his window down, squinting toward the tractor. He waved me down and called, "Sonny, how in the world is that truck staying so close to you?"

I laughed and looked back. From where he was sitting, it must've looked strange; a driverless truck trailing close behind my planter, keeping perfect pace. What he couldn't see was that inside that truck, barely tall enough to peek over the steering wheel, sat my eight-year-old son, Scott. His hands gripping the wheel, his chin just clearing the dash, his feet stretched out as far as they could reach the pedals.

"Don't worry," I shouted. "I've got a good driver in there!"

He just shook his head, smiling. And I still smile every time I think about that moment: the look on his face, the sun setting, and my boy driving that old truck like he'd been born to do it.

That's farming for you. It's a family affair; it always has been and always will be.

I was even younger than Scott the first time I got behind the wheel of that F-20 tractor. But that's how we learned. That's how life worked on the farm. Everyone pitched in. Everyone mattered. You didn't have to be grown-up to be useful.

Back then, that was just normal life. You worked alongside your parents, your grandparents, your brothers and sisters. And when Glendia and I started raising our own kids, it was the same way. We worked together because that's what families did. Farming wasn't just a job you clocked in and out of. It was your life; your reason to get up every morning.

It still amazes me that even today, with all the technology, all the pressure, and all the noise in the world, that family connection is still at the heart of farming. I don't think most people realize what a gift that is. How special it is to get to live and work side by side every single day.

Kids who grow up on farms see things most kids never do. They see their parents working hard, not for a paycheck that comes two weeks later, but for results that show up in the bin, in the barn, in the balance sheet when the crop's in. They see cause and effect in real time: a direct connection between sweat and success.

They see problem-solving in action. When something breaks, you fix it. When the weather turns, you adjust. When the odds are against you, you keep going. And through it all, they see what it means to work together. That's not just about chores, it's about trust. Parents teaching their kids how to hook up an implement or feed calves isn't just passing along responsibility. They're instilling independence and pride.

It's hard work, yes. It can be muddy, noisy, and exhausting. But it's also fun. What kid wouldn't jump at the chance to drive a truck or steer a tractor across a field? I remember when I was a boy, the kids from town were jealous of me. They thought driving at five years old was the wildest, freest thing in the world, and it was. I think that's part of the magic of growing up on a farm. You don't just watch life happen; you get to be a part of it.

A lot of people who don't know much about farming get it wrong. They think it's quiet, slow, maybe even boring. They picture isolation. But that's not what I've seen. Farm life is full of motion, energy, and laughter. Kids chasing calves, helping load wagons, running to the house when mom yells that supper's ready. It's a life full of purpose and love.

Today, there's a whole new dynamic to it. Farms are wired, connected, and computerized. Kids who grow up with smartphones and tablets are helping their parents track yields, calibrate planters, and manage livestock records. They're teaching us as much as we're teaching them. It's a new kind of family teamwork.

It's not just the farm, though. I think back to the small towns around us when I was growing up. The hardware stores, the feed mills, the corner groceries, they were all family-run. Moms and dads worked side by side, their kids sweeping floors, ringing up customers, or stocking shelves. That sense of ownership, that pride in building something together, ran deep.

Today, that's a lot harder to find. The number of kids growing up on farms or working in family businesses continues to decline. Parents head off to work away from home. Kids go to school, come home, and disappear into their rooms with their screens. The hours that used to be spent side by side are filled with something else: television, social media, video games, endless entertainment. It's no wonder families feel disconnected.

That gap worries me. Because when kids don't see their parents work, they miss out on something vital…not just how to work, but why we work. They miss the pride that comes from doing something real together.

EPILOGUE

I understand what people mean when they say we're richer now. We've got more conveniences, more gadgets, more options. But I don't think all that automatically adds up to being happier.

If you ask me, a big piece of the answer is right there on the family farm. It's in the rhythm of shared effort. Of getting up before sunrise, of getting dirty and working through the heat of the day. Of seeing what you've accomplished together when the sun finally sets.

Sometimes, when I see one of my great-grandkids riding in the seat of a tractor with their dad, I see that same spark I had when I was five. It reminds me that the heartbeat of a farm isn't in the soil or the seed, it's in the family that keeps showing up, generation after generation.

At Beck's, we've always believed that success never sleeps; not because we chase it endlessly, but because we live it every day through our purpose, our people, and our promise to stay family-owned and independent. The lessons learned in fields; perseverance, faith, humility, and hard work, are the same lessons that came from working side by side with family. Those same lessons built this company and will carry it forward for generations.

Sonny, Glendia, Tony, Kim, and Scott Beck (1975)

Scott, Francis, Sonny, and Tony Beck (1978)

Sonny and Scott Beck (1965)

Sonny and Kim (Beck) Marschand (1969)

Scott Beck on a Tractor (1970)

Sonny Pulling Tony Beck on a Sled With a Tractor (1973)

Tony, Kim, and Scott Beck on the First Day of School (1979)

Back to School Photo of Kim, Tony, and Scott Beck as Teenagers

Kim, Sonny, Glendia, Tony and Scott (1985)

Kim (Beck) Marschand Working in Customer Service and Computer Word Processing (1990)

Scott Beck in Early Days of His Beck's Career Managing the PFR Database, Advertising, and Seed Deliveries (1990)

Tony Beck Scouting Plots (1990)

Michelle Woods (Cousin of Scott, Kim, and Tony), Glendia Beck, Emily Marschand, and Kim (Beck) Marschand Operating Beck's Country Store

Tony Beck in a Field at Beck's East Central Illinois Production Farm That He Manages, Also Known as the Beck's Ranch (2024)

Sonny and Glendia in the Church Choir

The Family Bible Which is On Display at Beck's History Museum in Atlanta, Indiana

Three Generations of Beck Men – Sonny, Tony, Francis, and Scott (1992)

Three Generations of Beck Men – Sonny, Scott, Corey, Chris, and Jason (2024)

Sonny and Glendia Beck With Their Grandkids Celebrating Easter

Sonny and Glendia Beck With Their Grandkids Celebrating Christmas Morning

Sonny and Glendia Beck With Their Grandkids in a Seed Warehouse

Hallie, Faith, and Hannah Beck Outside the Beck Ranch in East Central Illinois.

The Family at the Beck Agricultural Center Ribbon Cutting at Purdue University (2005)

The Beck Grandkids on the Family's Annual Trip to Dale Hallow Lake in Tennessee

Sonny, Glendia, and Family in Front of Sonny's Restored Farmall F-20 in Celebration of Beck's 75th Anniversary (2011)

Bethany Gremel Interviewing Corey, Chris, and Jason Beck at the 2025 State of the Company Meeting

The Beck Family at the Indiana Living Legends Gala (2023)

Four Generations of the Beck Family Celebrating Christmas (2025)

The Beck Family at the 2026 Grand Opening of Purdue University's "The Lawrence," a Farm-To-Table Restaurant Housed in the Renovated Historic Beck Barn, Which Was Relocated To Purdue's Campus

Sonny Beck Sitting Down for an Interview for Beck's Home-Grown Legacy Documentary (2024)

DISCOVER BECK'S HOME-GROWN LEGACY

This window into the heart of Beck's is an inside look into what it takes to build something that lasts—not just a company, but a calling.

Still family-owned. Still independent. And still led by our family, who believe that loving what you do—and who you do it for—is the greatest privilege there is.

Family tradition got us here.
Hope for the future and faith in God keep us going.
And we can't wait to serve dedicated American farmers for generations to come.